INTERNET THEORY, TECHNOLOGY AND APPLICATIONS

COMPUTER NETWORKS: LEGAL POLICIES, PROGRAMS AND ISSUES

INTERNET THEORY, TECHNOLOGY AND APPLICATIONS

Additional books in this series can be found on Nova's website under the Series tab.

Additional E-books in this series can be found on Nova's website under the E-books tab.

LAWS AND LEGISLATION

Additional books in this series can be found on Nova's website under the Series tab.

Additional E-books in this series can be found on Nova's website under the E-books tab.

INTERNET THEORY, TECHNOLOGY AND APPLICATIONS

COMPUTER NETWORKS: LEGAL POLICIES, PROGRAMS AND ISSUES

RAYMOND S. GRIGELLO
EDITOR

Nova Science Publishers, Inc.
New York

For permission to use material from this book please contact us:
Telephone 631-231-7269; Fax 631-231-8175
Web Site: http://www.novapublishers.com

NOTICE TO THE READER

LIBRARY OF CONGRESS CATALOGING-IN-PUBLICATION DATA

Computer networks : legal policies, programs and issues / editor, Raymond S. Grigello.
p. cm.
Includes index.
ISBN 978-1-61209-596-7 (hardcover)
1. Computer networks--Law and legislation--United States. I. Grigello., Raymond S.
KF390.5.C6C65543 2011
343.09'944--dc22
2010054278

Published by Nova Science Publishers, Inc. † New York

CONTENTS

Preface		**vii**
Chapter 1	Net Neutrality: Background and Issues *Angele A. Gilroy*	**1**
Chapter 2	Broadband Internet Access and the Digital Divide: Federal Assistance Programs *Lennard G. Kruger and Angele A. Gilroy*	**11**
Chapter 3	Broadband Infrastructure Programs in the American Recovery and Reinvestment Act *Lennard G. Kruger*	**53**
Chapter 4	Comprehensive National Cybersecurity Initiative: Legal Authorities and Policy Considerations *John Rollins and Anna C. Henning*	**71**
Chapter 5	Spyware: Background and Policy Issues for Congress *Patricia Moloney Figliola*	**99**
Chapter 6	The Google Library Project: Is Digitization for Purposes of Online Indexing Fair Use under Copyright Law? *Kate M. Manuel*	**111**
Index		**129**

PREFACE

This book presents and discusses current legal policies and programs in place in regard to computer networks, as well as various issues which affect these procedures. Topics discussed include background and issues concerning net neutrality and spyware; broadband internet access and the digital divide; broadband infrastructure programs in the American Recovery and Reinvestment Act; the Comprehensive National Cybersecurity Initiative and the Google Library Project.

Chapter 1 – As congressional policymakers continue to debate telecommunications reform, a major point of contention is the question of whether action is needed to ensure unfettered access to the Internet. The move to place restrictions on the owners of the networks that compose and provide access to the Internet, to ensure equal access and non-discriminatory treatment, is referred to as "net neutrality."

There is no single accepted definition of "net neutrality." However, most agree that any such definition should include the general principles that owners of the networks that compose and provide access to the Internet should not control how consumers lawfully use that network; and should not be able to discriminate against content provider access to that network. Concern over whether it is necessary to take steps to ensure access to the Internet for content, services, and applications providers, as well as consumers, and if so, what these should be, is a major focus in the debate over telecommunications reform.

Some policymakers contend that more specific regulatory guidelines may be necessary to protect the marketplace from potential abuses which could threaten the net neutrality concept. Others contend that existing laws and Federal Communications Commission (FCC) policies are sufficient to deal with potential anticompetitive behavior and that such regulations would have negative effects on the expansion and future development of the Internet.

A consensus on this issue has not yet formed, and the 111th Congress, to date, has not introduced stand-alone legislation to address this issue. However, the net neutrality issue has been narrowly addressed within the context of the economic stimulus package (P.L. 111-5). Provisions in that law require the National Telecommunications and Information Administration (NTIA), in consultation with the FCC, to establish " ... nondiscrimination and network interconnection obligations" as a requirement for grant participants in the Broadband Technology Opportunities Program (BTOP).

Chapter 2 – The "digital divide" is a term that has been used to characterize a gap between "information haves and have-nots," or in other words, between those Americans who use or have access to telecommunications technologies (e.g., telephones, computers, the Internet) and those who do not. One important subset of the digital divide debate concerns high-speed Internet access and advanced telecommunications services, also known as broadband. Broadband is provided by a series of technologies (e.g., cable, telephone wire, fiber, satellite, wireless) that give users the ability to send and receive data at volumes and speeds far greater than traditional "dial-up" Internet access over telephone lines.

Broadband technologies are currently being deployed primarily by the private sector throughout the United States. While the numbers of new broadband subscribers continue to grow, studies and data suggest that the rate of broadband deployment in urban and high income areas are outpacing deployment in rural and low-income areas. Some policymakers, believing that disparities in broadband access across American society could have adverse economic and social consequences on those left behind, assert that the federal government should play a more active role to avoid a "digital divide" in broadband access. One approach is for the federal government to provide financial assistance to support broadband deployment in unserved and underserved areas.

Economic stimulus legislation enacted by the 111th Congress includes provisions that provides federal financial assistance for broadband deployment. On February 17, 2009, President Obama signed P.L. 111-5, the American Recovery and Reinvestment Act (ARRA). The ARRA provides a total of $7.2 billion for broadband, consisting of $4.7 billion to NTIA/DOC for a newly established Broadband Technology Opportunities Program and $2.5 billion to existing RUS/USDA broadband programs.

Meanwhile, it is expected that the Obama Administration will ultimately develop a national broadband policy or strategy that will seek to reduce or eliminate the "digital divide" with respect to broadband. It is likely that elements of a national broadband policy, in tandem with broadband investment measures in the American Recovery and Reinvestment Act, will significantly shape and expand federal policies and programs to promote broadband deployment and adoption. A key issue is how to strike a balance between providing federal assistance for unserved and underserved areas where the private sector may not be providing acceptable levels of broadband service, while at the same time minimizing any deleterious effects that government intervention in the marketplace may have on competition and private sector investment

Chapter 3 – The American Recovery and Reinvestment Act (ARRA, P.L. 111-5) provides $7.2 billion primarily for broadband grant programs to be administered by two separate agencies: the National Telecommunications and Information Administration (NTIA) of the Department of Commerce (DOC) and the Rural Utilities Service (RUS) of the U.S. Department of Agriculture (USDA). Of the $7.2 billion total, the ARRA provides $4.7 billion to establish a Broadband Technology Opportunities Program (BTOP) at NTIA, and $2.5 billion as additional funding for broadband grant, loan, and loan guarantee programs at RUS. Broadband grants and loans funded by the ARRA are competitive and applicants must apply directly to NTIA and RUS. The NTIA appropriation also includes $350 million for a national broadband inventory map, funding for the Broadband Data Improvement Act (P.L. 110-385), and funding to be transferred to the Federal Communications Commission (FCC) to develop a national broadband plan.

The unprecedented scale and scope of the ARRA broadband programs, coupled with the short time frame for awarding grants, presents daunting challenges with respect to program implementation as well as Congressional oversight. Congress is closely monitoring how equitably and effectively broadband grants are allocated among states and the various stakeholders, and to what extent the programs fulfill the goals of short term job creation and the longer term economic benefits anticipated from improved broadband availability, access, and adoption. A continuing issue is how to strike a balance between providing federal assistance for unserved and underserved areas where the private sector may not be providing acceptable levels of broadband service, while at the same time minimizing any deleterious effects that government intervention in the marketplace may have on competition and private sector investment.

Implementation decisions made by NTIA and RUS could have a significant impact on how the program is shaped and targeted, and the extent to which the program meets the goals and purposes set forth by the ARRA. Some implementation issues expected to be addressed include defining "underserved" and "unserved" areas with respect to broadband service, defining "nondiscrimination and network interconnection obligations," defining "broadband," the role of the states, coordination between federal agencies, broadband data collection, and evaluation and transparency.

Chapter 4 – Federal agencies report increasing cyber-intrusions into government computer networks, perpetrated by a range of known and unknown actors. In response, the President, legislators, experts, and others have characterized cybersecurity as a pressing national security issue.

Like other national security challenges in the post-9/11 era, the cyber threat is multi-faceted and lacks clearly delineated boundaries. Some cyber attackers operate through foreign nations' military or intelligence-gathering operations, whereas others have connections to terrorist groups or operate as individuals. Some cyber threats might be viewed as international or domestic criminal enterprises.

In January 2008, the Bush Administration established the Comprehensive National Cybersecurity Initiative (the CNCI) by a classified joint presidential directive. The CNCI establishes a multi- pronged approach the federal government is to take in identifying current and emerging cyber threats, shoring up current and future telecommunications and cyber vulnerabilities, and responding to or proactively addressing entities that wish to steal or manipulate protected data on secure federal systems. On February 9, 2009, President Obama initiated a 60-day interagency cybersecurity review to develop a strategic framework to ensure the CNCI is being appropriately integrated, resourced, and coordinated with Congress and the private sector.

In response to the CNCI and other proposals, questions have emerged regarding: (1) the adequacy of existing legal authorities—statutory or constitutional—for responding to cyber threats; and (2) the appropriate roles for the executive and legislative branches in addressing cybersecurity. The new and emerging nature of cyber threats complicates these questions. Although existing statutory provisions might authorize some modest actions, inherent constitutional powers currently provide the most plausible legal basis for many potential executive responses to national security related cyber incidences. Given that cyber threats originate from various sources, it is difficult to determine whether actions to prevent cyber attacks fit within the traditional scope of executive power to conduct war and foreign affairs.

Nonetheless, under the Supreme Court jurisprudence, it appears that the President is not prevented from taking action in the cybersecurity arena, at least until Congress takes further action. Regardless, Congress has a continuing oversight and appropriations role. In addition, potential government responses could be limited by individuals' constitutional rights or international laws of war. This report discusses the legal issues and addresses policy considerations related to the CNCI.

Chapter 5 – The term "spyware" generally refers to any software that is downloaded onto a computer without the owner's or user's knowledge. Spyware may collect information about a computer user's activities and transmit that information to someone else. It may change computer settings, or cause "pop-up" advertisements to appear (in that context, it is called "adware"). Spyware may redirect a Web browser to a site different from what the user intended to visit, or change the user's home page. A type of spyware called "keylogging" software records individual keystrokes, even if the author modifies or deletes what was written, or if the characters do not appear on the monitor. Thus, passwords, credit card numbers, and other personally identifiable information may be captured and relayed to unauthorized recipients.

Some of these software programs have legitimate applications the computer user wants. They obtain the moniker "spyware" when they are installed surreptitiously, or perform additional functions of which the user is unaware. Users typically do not realize that spyware is on their computer. They may have unknowingly downloaded it from the Internet by clicking within a website, or it might have been included in an attachment to an electronic mail message (e-mail) or embedded in other software.

The Federal Trade Commission (FTC) issued a consumer alert on spyware in October 2004. It provided a list of warning signs that might indicate that a computer is infected with spyware, and advice on what to do if it is.

Several states have passed spyware laws, but there is no specific federal law. Thus far, two bills have been introduced in the House of Representatives (H.R. 964 and H.R. 1525) and one has been introduced in the Senate (S. 1625). Both of the House bills have been reported and referred to the Senate. The Senate bill was referred to committee and no further action has been taken.

Chapter 6 – The Google Book Search Library Project, announced in December 2004, raised important questions about infringing reproduction and fair use under copyright law. Google planned to digitize, index, and display "snippets" of print books in the collections of five major libraries without the permission of the books' copyright holders, if any. Authors and publishers owning copyrights to these books sued Google in September and October 2005, seeking to enjoin and recover damages for Google's alleged infringement of their exclusive rights to reproduce and publicly display their works. Google and proponents of its Library Project disputed these allegations. They essentially contended that Google's proposed uses were not infringing because Google allowed rights holders to "opt out" of having their books digitized or indexed. They also argued that, even if Google's proposed uses were infringing, they constituted fair uses under copyright law.

The arguments of the parties and their supporters highlighted several questions of first impression. First, does an entity conducting an unauthorized digitization and indexing project avoid committing copyright infringement by offering rights holders the opportunity to "opt out," or request removal or exclusion of their content? Is requiring rights holders to take steps to stop allegedly infringing digitization and indexing like requiring rights holders to use meta-tags to keep search engines from indexing online content? Or do rights holders employ sufficient measures to keep their books from being digitized and indexed online by publishing in print? Second, can unauthorized digitization, indexing, and display of "snippets" of print works constitute a fair use? Assuming unauthorized indexing and display of "snippets" are fair uses, can digitization claim to be a fair use on the grounds that apparently prima facie infringing activities that facilitate legitimate uses are fair uses?

On October 28, 2008, Google, authors, and publishers announced a proposed settlement, which, if approved by the court, could leave these and related questions unanswered. However, although a court granted preliminary approval to the settlement on November 17, 2008, final approval is still pending. Until final approval is granted, any rights holder belonging to the proposed settlement class—which includes "all persons having copyright interests in books" in the United States— could object to the agreement. The court could also reject the agreement as unfair, unreasonable, or inadequate. Moreover, even assuming final court approval, future cases may raise similar questions about infringing reproduction and fair use.

Versions of these chapters were also published in *Internet Journal,* Volume 2 Issue 4, published by Nova Science Publishers, Inc. They were submitted for appropriate modifications in an effort to encourage wider dissemination of research.

In: Computer Networks
Editor: Raymond S. Grigello

ISBN: 978-1-61209-596-7

Chapter 1

NET NEUTRALITY: BACKGROUND AND ISSUES*

Angele A. Gilroy#

ABSTRACT

As congressional policymakers continue to debate telecommunications reform, a major point of contention is the question of whether action is needed to ensure unfettered access to the Internet. The move to place restrictions on the owners of the networks that compose and provide access to the Internet, to ensure equal access and non-discriminatory treatment, is referred to as "net neutrality."

There is no single accepted definition of "net neutrality." However, most agree that any such definition should include the general principles that owners of the networks that compose and provide access to the Internet should not control how consumers lawfully use that network; and should not be able to discriminate against content provider access to that network. Concern over whether it is necessary to take steps to ensure access to the Internet for content, services, and applications providers, as well as consumers, and if so, what these should be, is a major focus in the debate over telecommunications reform.

Some policymakers contend that more specific regulatory guidelines may be necessary to protect the marketplace from potential abuses which could threaten the net neutrality concept. Others contend that existing

* This is an edited, excerpted and augmented edition of a CRS Report, Report RS22444 dated March 19, 2009.

laws and Federal Communications Commission (FCC) policies are sufficient to deal with potential anticompetitive behavior and that such regulations would have negative effects on the expansion and future development of the Internet.

A consensus on this issue has not yet formed, and the 111th Congress, to date, has not introduced stand-alone legislation to address this issue. However, the net neutrality issue has been narrowly addressed within the context of the economic stimulus package (P.L. 111-5). Provisions in that law require the National Telecommunications and Information Administration (NTIA), in consultation with the FCC, to establish " ... nondiscrimination and network interconnection obligations" as a requirement for grant participants in the Broadband Technology Opportunities Program (BTOP).

NETWORK NEUTRALITY

As congressional policymakers continue to debate telecommunications reform, a major point of contention is the question of whether action is needed to ensure unfettered access to the Internet. The move to place restrictions on the owners of the networks that compose and provide access to the Internet, to ensure equal access and non-discriminatory treatment, is referred to as "net neutrality." There is no single accepted definition of "net neutrality." However, most agree that any such definition should include the general principles that owners of the networks that compose and provide access to the Internet should not control how consumers lawfully use that network; and should not be able to discriminate against content provider access to that network.

What, if any, action should be taken to ensure "net neutrality" has become a major focal point in the debate over broadband, or high-speed Internet access, regulation. As the marketplace for broadband continues to evolve, some contend that no new regulations are needed, and if enacted will slow deployment of and access to the Internet, as well as limit innovation. Others, however, contend that the consolidation and diversification of broadband providers into content providers has the potential to lead to discriminatory behaviors which conflict with net neutrality principles. The two potential behaviors most often cited are the network providers' ability to control access to and the pricing of broadband facilities, and the incentive to favor network-owned content, thereby placing unaffiliated content providers at a competitive disadvantage [1].

In 2005 two major actions dramatically changed the regulatory landscape as it applied to broadband services, further fueling the net neutrality debate. In both cases these actions led to the classification of broadband Internet access services as Title I information services, thereby subjecting them to a less rigorous regulatory framework than those services classified as telecommunications services. In the first action, the U.S. Supreme Court, in a June 2005 decision (*National Cable & Telecommunications Association v. Brand X Internet Services*), upheld the Federal Communications Commission's (FCC) 2002 ruling that the provision of cable modem service (i.e., cable television broadband Internet) is an interstate information service and is therefore subject to the less stringent regulatory regime under Title I of the Communications Act of 1934.[2] In a second action, the FCC in an August 5, 2005 decision, extended the same regulatory relief to telephone company Internet access services (i.e., wireline broadband Internet access, or DSL), thereby also defining such services as information services subject to Title I regulation [3]. As a result neither telephone companies nor cable companies, when providing broadband services, are required to adhere to the more stringent regulatory regime for telecommunications services found under Title II (common carrier) of the 1934 Act. [4]. However, classification as an information service does not free the service from regulation. The FCC continues to have regulatory authority over information services under its Title I, ancillary jurisdiction [5].

Simultaneous to the issuing of its August 2005 information services classification order, the FCC also adopted a policy statement outlining the following four principles to "encourage broadband deployment and preserve and promote the open and interconnected nature of [the] public Internet:" (1) consumers are entitled to access the lawful Internet content of their choice; (2) consumers are entitled to run applications and services of their choice (subject to the needs of law enforcement); (3) consumers are entitled to connect their choice of legal devices that do not harm the network; and (4) consumers are entitled to competition among network providers, application and service providers, and content providers. Then FCC Chairman Martin did not call for their codification. However, he stated that they will be incorporated into the policymaking activities of the Commission.[6] For example, one of the agreed upon conditions for the October 2005 approval of both the Verizon/MCI and the SBC/AT&T mergers was an agreement made by the involved parties to commit, for two years, "... to conduct business in a way that comports with the Commission's (September 2005) Internet policy statement.... " [7]. In a further action AT&T included in its concessions to gain FCC approval of its merger to BellSouth to adhering, for two years, to significant net neutrality requirements. Under terms of the merger agreement, which was approved on December 29, 2006, AT&T agreed to not only uphold, for 30 months, the FCC's Internet policy statement principles, but also committed, for two years (expired December 2008), to stringent requirements to "... maintain a neutral network and neutral routing in its wireline broadband Internet access service." [8]

In perhaps one of its most significant actions relating to its Internet policy statement to date, the FCC, on August 1, 2008, ruled that Comcast Corp., a provider of Internet access over cable lines, violated the FCC's policy statement, when it selectively blocked peer-to-peer connections in an attempt to manage its traffic [9] This practice, the FCC concluded, "... unduly interfered with Internet users' rights to access the lawful Internet content and to use the applications of their choice." While no monetary penalties were imposed, Comcast is required to stop these practices by the end of 2008. Comcast stated that it will comply with the order, but it has filed an appeal in the U.S. DC Court of Appeals [10].

Separately, in an April 2007 action, the FCC released a notice of inquiry (WC Docket No. 07-52), which is still pending, on broadband industry practices seeking comment on a wide range of issues including whether the August 2005 Internet policy statement should be amended to incorporate a new principle of nondiscrimination and if so, what form it should take. On January 14, 2008 the FCC issued three public notices seeking comment on issues related to network management (including the now-completed Comcast ruling) and held two (February 25 and April 17, 2008) public hearings specific to broadband network management practices.

Network Prioritization

As consumers expand their use of the Internet and new multimedia and voice services become more commonplace, control over network quality also becomes an issue. In the past, Internet traffic has been delivered on a "best efforts" basis. The quality of service needed for the delivery of the most popular uses, such as email or surfing the Web, is not as dependent on guaranteed quality. However, as Internet use expands to include video, online gaming, and voice service, the need for uninterrupted streams of data becomes important. As the demand for such services continues to expand, network broadband operators are moving to prioritize network traffic to ensure the quality of these services. Prioritization may benefit consumers by ensuring faster delivery and quality of service and may be necessary to ensure the proper functioning of expanded service options. However, the move on the part of network operators to establish prioritized networks, while embraced by some, has led to a number of policy concerns.

There is concern that the ability of network providers to prioritize traffic may give them too much power over the operation of and access to the Internet. If a multi-tiered Internet develops where content providers pay for different service levels, the potential to limit competition exists, if smaller, less financially secure content providers are unable to afford to pay for a higher level of access. Also, if network providers have control over who is given priority access, the ability to discriminate among who gets such access is also present. If such a scenario were to develop, the potential benefits to consumers of a prioritized network would be lessened by a decrease in consumer choice and/or increased costs, if the fees charged for premium access are passed on to the consumer. The potential for these abuses, however, is significantly decreased in a marketplace where multiple, competing broadband providers exist. If a network broadband provider blocks access to content or charges unreasonable fees, in a competitive market, content providers and consumers could obtain their access from other network providers. As consumers and content providers migrate to competitors, market share and profits of the offending network provider will decrease leading to corrective action or failure. However, this scenario assumes that every market will have a number of equally competitive broadband options from which to choose, and all competitors will have equal access to, if not identical, at least comparable content.

Despite the FCC's ability to regulate broadband services under its Title I ancillary authority and the issuing of its broadband principles, some policymakers feel that more specific regulatory guidelines may be necessary to protect the marketplace from potential abuses; a consensus on what these should specifically entail, however, has yet to form. Others feel that existing laws and FCC policies regarding competitive behavior are sufficient to deal with potential anti-competitive behavior and that no action is needed and if enacted at this time, could result in harm.

The Congressional Debate

The issue of net neutrality, and whether legislation is needed to ensure access to broadband networks and services, has become a major focal point in the debate over telecommunications reform [11]. Those opposed to the enactment of legislation to impose specific Internet network access or "net neutrality" mandates claim that such action goes against the long standing policy to keep the Internet as free as possible from regulation. The imposition of such requirements, they state, is not only unnecessary, but would have negative consequences for the deployment and advancement of broadband facilities. For example, further expansion of networks by existing providers and the entrance of new network providers, would be discouraged, they claim, as investors would be less willing to finance networks that may be operating under mandatory build- out and/or access requirements. Application innovation could also be discouraged, they contend, if, for example, network providers are restricted in the way they manage their networks or are limited in their ability to offer new service packages or formats. Such legislation is not needed, they claim, as major Internet access providers have stated publicly that they are committed to upholding the FCC's four policy principles [12]. Opponents also state that advocates of regulation cannot point to any widespread behavior that justifies the need to establish such regulations and note that competition between telephone and cable system providers, as well as the growing presence of new technologies (e.g., satellite, wireless, and power lines) will serve to counteract any potential anti-discriminatory behavior. Furthermore, opponents claim, even if such a violation should occur, the FCC already has the needed authority to pursue violators. They note that the FCC has not requested further authority [13] and has successfully used its existing authority, in the August 1, 2008, Comcast decision (see above) as well as in a March 3, 2005, action against Madison River Communications. In the latter case, the FCC intervened and resolved, through a consent decree, an alleged case of port blocking by Madison River Communications, a local exchange (telephone) company [14]. The full force of antitrust law is also available, they claim, in cases of discriminatory behavior.

Proponents of net neutrality legislation, however, feel that absent some regulation, Internet access providers will become gatekeepers and use their market power to the disadvantage of Internet users and competing content and application providers. They cite concerns that the Internet could develop into a two-tiered system favoring large, established businesses or those with ties to broadband network providers. While market forces should be a deterrent to such anti-competitive behavior, they point out that today's market for residential broadband delivery is largely dominated by only two providers, the telephone and cable television companies, and that, at a minimum, a strong third player is needed to ensure that the benefits of competition will prevail [15]. The need to formulate a national policy to clarify expectations and ensure the "openness" of the Internet is important to protect the benefits and promote the further expansion of broadband, they claim. The adoption of a single, coherent, regulatory framework to prevent discrimination, supporters claim, would be a positive step for further development of the Internet, by providing the marketplace stability needed to encourage investment and foster the growth of new services and applications. Furthermore, relying on current laws and case-by-case anti-trust-like enforcement, they claim, is too cumbersome, slow, and expensive, particularly for small start-up enterprises [16].

Congressional Activity

The 110th Congress addressed the debate over net neutrality largely within the broader issue of telecommunications reform. Then House Telecommunications and the Internet Subcommittee Chairman Markey, a strong advocate of net neutrality legislation, introduced legislation (H.R. 5353) to address this issue and held a May 6, 2008 hearing on the measure. House Judiciary Chairman Conyers introduced H.R. 5994, a bill which establishes an antitrust approach to address anticompetitive and discriminatory practices by broadband providers as a follow-up to a March 11, 2008 hearing on net neutrality held by the House Judiciary Antitrust Task Force. A standalone net neutrality measure (S. 215) was introduced and referred to the Senate Commerce, Science, and Transportation Committee where an April 22, 2008 hearing on the "Future of the Internet" was held. No further activity was undertaken in the 110th Congress.

A consensus on this issue has not yet formed, and no stand-alone measures addressing net neutrality have been introduced in the 111th Congress, to date. House Communications, Technology, and the Internet Subcommittee Chairman Boucher has stated that he continues to work with broadband providers and content providers to seek common ground on network management practices, and at this time, is pursuing this approach.

However, the net neutrality issue has been narrowly addressed within the context of the economic stimulus package. H.R. 1 (P.L. 111-5) contains provisions that require the National Telecommunications and Information Administration (NTIA), in consultation with the FCC, to establish "... nondiscrimination and network interconnection obligations" as a requirement for grant participants in the Broadband Technology Opportunities Program (BTOP). The law further directs that the FCC's four broadband policy principles, issued in August 2005, are the minimum obligations to be imposed [17]. The NTIA has not, as of yet, issued these requirements.

REFERENCES

[1] The practice of consumer tiering, that is the charging of different rates to subscribers based on access speed, is not the concern. Access tiering, that is the charging of different fees, or the establishment of different terms and conditions to content, services, or applications providers for access to the broadband pipe is the focus of the net neutrality policy debate.

[2] 47U.S.C. 151 et seq. For a full discussion of the Brand X decision see CRS Report RL32985, *Defining Cable Broadband Internet Access Service: Background and Analysis of the Supreme Court's Brand X Decision*, by Angie A. Welborn and Charles B. Goldfarb.

[3] Seehttp://hraunfoss.fcc.gov/edocs_public/attachmatch/DOC-2604 33A2.pdf for a copy of FCC Chairman Martin's statement. For a summary of the final rule see Appropriate Framework for Broadband Access to the Internet Over Wireline Facilities. *Federal Register*, Vol. 70, No. 199, October, 17, 2005, p. 60222.

[4] For example, Title II regulations impose rigorous anti-discrimination, interconnection and access requirements. For a further discussion of Title I versus Title II regulatory authority see CRS Report RL32985, cited above.

[5] Title I of the 1934 Communications Act gives the FCC such authority if assertion of jurisdiction is "reasonably ancillary to the effective performance of [its] various responsibilities." The FCC in its order, cites consumer protection, network reliability, or national security obligations as examples of cases where such authority would apply (see paragraph 36 of the final rule summarized in the *Federal Register* cite in footnote 3, above).

[6] See http://www.fcc.gov/headlines2005.html. August 5, 2005. *FCC Adopts Policy Statement on Broadband Internet Access*.

[7] See: http://hraunfoss.FCC.gov/edocs_public/attachmatch/DOC-261936A1.pdf. It should be noted that applicants offered certain voluntary commitments, of which this was one.

[8] See: http://hraunfoss.fcc.gov/edocs_public/attachmatch/DOC-269275A1.pdf.

[9] See: http://hraunfoss.fcc.gov/edocs_public/attachmatch/FCC-08-183A1.pdf.

[10] For a legal discussion of the FCC's Comcast decision see CRS Report R40234, *Net Neutrality: The Federal Communications Commission's Authority to Enforce Its Network Management Principles* , by Kathleen Ann Ruane.

[11] For a more lengthy discussion regarding proponents' and opponents' views see, for example, testimony from Senate Commerce Committee hearings on Net Neutrality, February 7, 2006; http://commerce.senate.gov/public/index.cfm? FuseAction=Hearings. Hearing&Hearing_ID= 1708.

[12] See testimony of Kyle McSlarrow, President and CEO of the National Cable and Telecommunications Association and Walter McCormick, President and CEO of the United States Telecom Association, hearing on Net Neutrality before the Senate Commerce Committee, February 7, 2006, cited above.

[13] Former FCC Chairman Martin indicated that the FCC has the necessary tools to uphold the FCC's stated policy principles and did not requested additional authority. Furthermore, former Chairman Martin stated that he was "... confident that the marketplace will continue to ensure that these principles are maintained" and is "... confident therefore, that regulation is not, nor will be, required." See former *Chairman Kevin J. Martin Comments on Commission Policy Statement*, at http://hraunfoss.fcc.gov/edocs_public/attachmatch/DOC-260435A2.pdf. However, FCC Commissioner Copps, in an April 3, 2006 speech, did express concerns over the concentration in broadband facilities providers and their " ... ability, and possibly even the incentive, to act as Internet gatekeepers ... " and called for a "national policy" on "... issues regarding consumer rights, Internet openess, and broadband deployment." See http://hraunfoss.fcc.gov/edocs_ public/attachmatch/DOC-264765A1.pdf, for a copy of Commissioner Copps' speech.

[14] The FCC entered into a consent decree with Madison River Communications to settle charges that the company had deliberately

blocked the ports on its network that were used by Vonage Corp. to provide voice over Internet protocol (VoIP) service. Under terms of the decree Madison River agreed to pay a $15,000 fine and not block ports used for VoIP applications. See http://hraunfoss.fcc.gov/edocs_public/attachmatch/DA-05-543A2.pdf. for a copy of the consent decree

[15] For FCC market share data for high-speed connections see *High-Speed Services for Internet Access: Status as of June 30, 2007,* Federal Communications Commission, Industry Analysis and Technology Division, Wireline Competition Bureau, released March 2008. View report at http://hraunfoss.fcc.gov/edocs_public/attachmatch/DOC2809 06A1 .pdf.

[16] For example, see testimony of Vint Cerf, VP Google, Earl Comstock, President and CEO of CompTel, and Jeffrey Citron, Chairman and CEO Vonage, hearing on Net Neutrality, before the Senate Commerce Committee, February 7, 2006, cited above.

[17] For a further more detailed discussion of the broadband infrastructure programs contained in P.L. 111-5 see CRS Report R40436, *Broadband Infrastructure Programs in the American Recovery and Reinvestment Act*, by Lennard G. Kruger.

In: Computer Networks
Editor: Raymond S. Grigello

ISBN: 978-1-61209-596-7

Chapter 2

BROADBAND INTERNET ACCESS AND THE DIGITAL DIVIDE: FEDERAL ASSISTANCE PROGRAMS*

Lennard G. Kruger and Angele A. Gilroy

ABSTRACT

The "digital divide" is a term that has been used to characterize a gap between "information haves and have-nots," or in other words, between those Americans who use or have access to telecommunications technologies (e.g., telephones, computers, the Internet) and those who do not. One important subset of the digital divide debate concerns high-speed Internet access and advanced telecommunications services, also known as *broadband.* Broadband is provided by a series of technologies (e.g., cable, telephone wire, fiber, satellite, wireless) that give users the ability to send and receive data at volumes and speeds far greater than traditional "dial-up" Internet access over telephone lines.

Broadband technologies are currently being deployed primarily by the private sector throughout the United States. While the numbers of new broadband subscribers continue to grow, studies and data suggest that the rate of broadband deployment in urban and high income areas are outpacing deployment in rural and low-income areas. Some policymakers, believing that disparities in broadband access across American society could have adverse economic and social consequences

* This is an edited, excerpted and augmented edition of a CRS Report, Report RL30719 dated March 19, 2009.

on those left behind, assert that the federal government should play a more active role to avoid a "digital divide" in broadband access. One approach is for the federal government to provide financial assistance to support broadband deployment in unserved and underserved areas.

Economic stimulus legislation enacted by the 111th Congress includes provisions that provides federal financial assistance for broadband deployment. On February 17, 2009, President Obama signed P.L. 111-5, the American Recovery and Reinvestment Act (ARRA). The ARRA provides a total of $7.2 billion for broadband, consisting of $4.7 billion to NTIA/DOC for a newly established Broadband Technology Opportunities Program and $2.5 billion to existing RUS/USDA broadband programs.

Meanwhile, it is expected that the Obama Administration will ultimately develop a national broadband policy or strategy that will seek to reduce or eliminate the "digital divide" with respect to broadband. It is likely that elements of a national broadband policy, in tandem with broadband investment measures in the American Recovery and Reinvestment Act, will significantly shape and expand federal policies and programs to promote broadband deployment and adoption. A key issue is how to strike a balance between providing federal assistance for unserved and underserved areas where the private sector may not be providing acceptable levels of broadband service, while at the same time minimizing any deleterious effects that government intervention in the marketplace may have on competition and private sector investment.

INTRODUCTION

The "digital divide" is a term used to describe a perceived gap between perceived "information haves and have-nots," or in other words, between those Americans who use or have access to telecommunications technologies (e.g., telephones, computers, the Internet) and those who do not [1]. Whether or not individuals or communities fall into the "information haves" category depends on a number of factors, ranging from the presence of computers in the home, to training and education, to the availability of affordable Internet access.

Broadband technologies are currently being deployed primarily by the private sector throughout the United States. While the numbers of new broadband subscribers continue to grow, studies and data suggest that the rate of broadband deployment in urban and high income areas are outpacing deployment in rural and low-income areas. Some policymakers, believing that disparities in broadband access across American society could have adverse economic and social consequences on those left behind, assert that the federal government should play a more active role to avoid a "digital divide" in broadband access. One approach—adopted in the American Recovery and Reinvestment Act of 2009 (P.L. 111-5)—is for the federal government to provide financial assistance, primarily grants, to support broadband deployment in unserved and underserved areas.

Status of Broadband Deployment in the United States

Prior to the late 1990s, American homes accessed the Internet at maximum speeds of 56 kilobits per second by dialing up an Internet Service Provider (such as AOL) over the same copper telephone line used for traditional voice service. A relatively small number of businesses and institutions used broadband or high speed connections through the installation of special "dedicated lines" typically provided by their local telephone company. Starting in the late 1990s, cable television companies began offering cable modem broadband service to homes and businesses. This was accompanied by telephone companies beginning to offer DSL service (broadband over existing copper telephone wireline). Growth has been steep, rising from 2.8 million high speed lines reported as of December 1999, to 121.2 million lines as of December 31, 2007. Of the 121.2 million high speed lines reported by the FCC, 74.0 million serve residential users [2]. Since the deployment of residential broadband in the United States, the primary residential broadband technologies deployed continue to be cable modem and DSL. A distinction is often made between "current generation" and "next generation" broadband (commonly referred to as next generation networks or NGN). "Current generation" typically refers to currently deployed cable, DSL, and many wireless systems, while "next generation" refers to dramatically faster download and upload speeds offered by fiber technologies and also potentially by future generations of cable, DSL, and wireless technologies [3]. In general, the greater the download and upload speeds offered by a broadband connection, the more sophisticated (and potentially valuable) the application that is enabled.

December 2008 survey data from the Pew Internet and American Life Project found that 57% of Americans have broadband at home [4]. It is estimated that less than 10% of U.S. households have no access to any broadband provider whatsoever (not including satellite) [5]. While the broadband *adoption* or *penetration* rate stands at close to 60% of U.S.

households, broadband *availability* is much higher, at more than 90% of households. Thus, approximately 30% of households have access to some type of terrestrial (non-satellite) broadband service, but do not choose to subscribe. According to the FCC, possible reasons for the gap between broadband availability and subscribership include the lack of computers in some homes, price of broadband service, lack of content, and the availability of broadband at work [6]. According to Pew, non-broadband users tend to be older, have lower incomes, have trouble using technology, and may not see the relevance of using the Internet to their lives. Between 2007 and 2008, low income Americans (under $20,000 annual income) and African Americans showed no significant growth in home broadband adoption after strong growth in previous years [7] Pew also found that about one-third of adults without broadband cite price and availability as the reasons why they don't have broadband in their homes, while two-thirds cite reasons such as usability and relevance [8].

Broadband in Rural and Underserved Areas [9]

While the number of new broadband subscribers continues to grow, the rate of broadband deployment in urban and high income areas appears to be outpacing deployment in rural and low-income areas. While there are many examples of rural communities with state of the art telecommunications facilities, [10] recent surveys and studies have indicated that, in general, rural areas tend to lag behind urban and suburban areas in broadband deployment. Data (2008) from the Pew Internet and American Life Project indicate that while broadband adoption is growing in urban, suburban, and rural areas, broadband users make up larger percentages of urban and suburban users than rural users. Pew found that the percentage of all U.S. adults with broadband at home is 60% for suburban areas, 57% for urban areas, and 38% for rural areas [11].

Similarly, according to the latest FCC data on the deployment of high-speed Internet connections (released January 2009), high-speed subscribers were reported in 99% of the most densely populated zip codes, as opposed to 90% of zip codes with the lowest population densities. For zip codes ranked by median family income, high-speed subscribers were reported present in 99% of the top one-tenth of zip codes, as compared to 92% of the bottom one-tenth of zip codes [12].

The comparatively lower population density of rural areas is likely the major reason why broadband is less deployed than in more highly populated suburban and urban areas. Particularly for wireline broadband technologies—such as cable modem and DSL—the greater the geographical distances among customers, the larger the cost to serve those customers. Thus, there is often less incentive for companies to invest in broadband in rural areas than, for example, in an urban area where there is more demand (more customers with perhaps higher incomes) and less cost to wire the market area [13].

Some policymakers believe that disparities in broadband access across American society could have adverse consequences on those left behind, and that advanced telecommunications applications critical for businesses and consumers to engage in e-commerce are increasingly dependent on high speed broadband connections to the Internet. Thus, some say, communities and individuals without access to broadband could be at risk to the extent that e-commerce becomes a critical factor in determining future economic development and prosperity. A February 2006 study done by the Massachusetts Institute of Technology for the Economic Development Administration of the Department of Commerce marked the first attempt to quantitatively measure the impact of broadband on economic growth. The study found that "between 1998 and 2002, communities in which mass-market broadband was available by December 1999 experienced more rapid growth in employment, the number of businesses overall, and businesses in IT-intensive sectors, relative to comparable communities without broadband at that time." [14].

Subsequently, a June 2007 report from the Brookings Institution found that for every one percentage point increase in broadband penetration in a state, employment is projected to increase by 0.2 to 0.3% per year. For the entire U.S. private non-farm economy, the study projected an increase of about 300,000 jobs [15].

Some also argue that broadband is an important contributor to U.S. future economic strength with respect to the rest of the world. According to the International Telecommunications Union, the U.S. ranks 24th worldwide in broadband penetration (subscriptions per 100 inhabitants in 2007) [16]. Data from the Organization for Economic Cooperation and Development (OECD) found the U.S. ranking 15th among OECD nations in broadband access per 100 inhabitants as of June 2008 [17]. By contrast, in 2001 an OECD study found the U.S. ranking 4th in broadband subscribership per 100 inhabitants (after Korea, Sweden, and Canada) [18]. While many argue that the U.S. declining performance in international broadband rankings is a cause for concern, [19] others maintain that the OECD and ITU data undercount U.S. broadband deployment, [20] and that cross-country broadband deployment comparisons are not necessarily meaningful and inherently problematic [21] Finally, an issue related to international broadband rankings is the extent to which broadband speeds and prices differ between the U.S. and the rest of the world [22].

Is Broadband Deployment Data Adequate?

Obtaining an accurate snapshot of the status of broadband deployment is problematic. Anecdotes abound of rural and low-income areas which do not have adequate Internet access, as well as those which are receiving access to high-speed, state-of-the-art connections. Rapidly evolving technologies, the constant flux of the telecommunications industry, the uncertainty of consumer wants and needs, and the sheer diversity and size of the nation's economy and geography make the status of broadband deployment very difficult to characterize. The FCC periodically collects broadband deployment data from the private sector via "FCC Form 477"—a standardized information gathering survey. Statistics derived from the Form 477 survey are published every six months. Additionally, data from Form 477 are used as the basis of the FCC's (to date) five broadband deployment reports.

The FCC is working to refine the data used in future Reports in order to provide an increasingly accurate portrayal. In its March 17, 2004 Notice of Inquiry for the *Fourth Report*, the FCC sought comments on specific proposals to improve the FCC Form 477 data gathering program [23]. On November 9, 2004, the FCC voted to expand its data collection program by requiring reports from all facilities based carriers regardless of size in order to better track rural and underserved markets, by requiring broadband providers to provide more information on the speed and nature of their service, and by establishing broadband-over-power line as a separate category in order to track its development and deployment. The FCC Form 477 data gathering program was extended for five years beyond its March 2005 expiration date [24].

The Government Accountability Office (GAO) has cited concerns about the FCC's zip-code level data. Of particular concern is that the FCC will report broadband service in a zip code even if a company reports service to only one subscriber, which in turn can lead to some observers overstating broadband deployment. According to GAO, "the data may not provide a highly accurate depiction of local deployment of broadband infrastructures for residential service, especially in rural areas." The FCC has acknowledged the limitations in its zip code level data [25].

On April 16, 2007, the FCC announced a Notice of Proposed Rulemaking which sought comment on a number of broadband data collection issues, including how to develop a more accurate picture of broadband deployment; gathering information on price, other factors determining consumer uptake of broadband, and international comparisons; how to improve data on wireless broadband; how to collect information on subscribership to voice over Internet Protocol service (VoIP); and whether to modify collection of speed tier information [26].

On March 19, 2008, the FCC adopted an Order that substantially expands its broadband data collection capability. Specifically, the Order expands the number of broadband reporting speed tiers to capture more information about upload and download speeds offered in the marketplace, requires broadband providers to report numbers of broadband subscribers by census tract, and improves the accuracy of information collected on mobile wireless broadband deployment. Additionally, in a Further Notice of Proposed Rulemaking, the FCC sought comment on broadband service pricing and availability [27].

During the 110th Congress, state initiatives to collect broadband deployment data in order to promote broadband in underserved areas were viewed as a possible model for governmental efforts to encourage broadband. In particular, an initiative in the Commonwealth of Kentucky— called ConnectKentucky—has developed detailed broadband inventory mapping which identifies local communities that lack adequate broadband service. Kentucky is using this data to promote public-private partnerships in order to reach a goal of universal broadband coverage in the state [28]. Other states are pursuing or considering similar approaches.

The 110th Congress explored ways to support or implement the types of broadband mapping and data collection efforts demonstrated by ConnectKentucky. The Broadband Data Improvement Act was enacted by the 110th Congress and became P.L. 110-385 on October 10, 2008. The law requires the FCC to collect demographic information on unserved areas, data comparing broadband service with 75 communities in at least 25 nations abroad, and data on consumer use of broadband. The act also directs the Census Bureau to collect broadband data, the Government Accountability Office to study broadband data metrics and standards, and the Department of Commerce to provide grants supporting state broadband initiatives.

P.L. 111-5, the American Recovery and Reinvestment Act, provides NTIA with an appropriation of $350 million to implement the Broadband Data Improvement Act and to develop and maintain a national broadband inventory map, which shall be made accessible to the public no later than two years after enactment.

BROADBAND AND THE FEDERAL ROLE

The Telecommunications Act of 1996 (P.L. 104-104) addressed the issue of whether the federal government should intervene to prevent a "digital divide" in broadband access. Section 706 requires the FCC to determine whether "advanced telecommunications capability [i.e., broadband or high-speed access] is being deployed to all Americans in a reasonable and timely fashion." If this is not the case, the act directs the FCC to "take immediate action to accelerate deployment of such capability by removing barriers to infrastructure investment and by promoting competition in the telecommunications market."

Since 1999, the FCC has issued and adopted five reports pursuant to Section 706. All five reports formally concluded that the deployment of advanced telecommunications capability to all Americans is reasonable and timely. The fifth and most recent 706 report was adopted on March 19, 2008, and released on June 12, 2008 [29]. Two FCC Commissioners (Michael Copps and Jonathan Adelstein) have repeatedly dissented from the reports' conclusions that broadband deployment is reasonable and timely, arguing that the relatively poor world ranking of United States broadband penetration indicates that deployment is insufficient, that the FCC's definition of broadband was outdated and not comparable to the much higher speeds available to consumers in other countries, that the use of zip code data (measuring the presence of at least one broadband subscriber within a zip code area) did not sufficiently characterize the availability of broadband across geographic areas, and that broadband deployment is impeded by the lack of a comprehensive national broadband policy [30].

Bush Administration

The Bush Administration pursued a broadband policy that emphasized deregulation, non-intervention by government in the marketplace, and general tax policies intended to foster overall economic growth. On March 26, 2004, President Bush endorsed a goal of "universal broadband access by 2007," and on April 26, 2004, announced a broadband initiative which included promoting legislation which would permanently prohibit all broadband taxes, making spectrum available for wireless broadband and creating technical standards for broadband over power lines, and simplifying rights-of-way processes on federal lands for broadband providers [31]. Subsequently, on January 31, 2008, NTIA released a report, entitled, *Networked Nation: Broadband in America, 2007* which characterized the Bush Administration's broadband initiative as follows:

> From its first days, the Administration has implemented a comprehensive and integrated package of technology, regulatory, and fiscal policies designed to lower barriers and create an environment in which broadband innovation and competition can flourish [32].

The Bush Administration broadband policy embraced the view that a minimum of government intervention would create an economic climate favorable to private sector investment in the broadband market. According to NTIA, the report showed "that the Administration's technology, regulatory, and fiscal policies have stimulated innovation and competition, and encouraged investment in the U.S. broadband market contributing to significantly increased accessibility of broadband services."[33]

During the 110th Congress, some policymakers disagreed with the Bush Administration's assessment and asserted that the federal government should play a more active role to avoid a "digital divide" in broadband access. Bills were introduced seeking to provide federal financial assistance for broadband deployment in the form of grants, loans, subsidies, and/or tax credits.

Obama Administration

It is expected that the Obama Administration will ultimately develop a national broadband policy or strategy that will seek to reduce or eliminate the "digital divide" with respect to broadband. One of the key elements of the Obama transition's technology agenda is to "deploy next-generation broadband," and specifically:

> Work towards true broadband in every community in America through a combination of reform of the Universal Service Fund, better use of the nation's wireless spectrum, promotion of next-generation facilities, technologies and applications, and new tax and loan incentives. America should lead the world in broadband penetration and Internet access [34].

The Obama campaign released a policy blueprint for technology and innovation that includes policy proposals intended to result in full broadband penetration and deployment of next-generation broadband. Specifically, policy proposals include:

- redefining broadband at speeds "demanded by 21st century business and communications;"
- reforming universal service to support affordable broadband specifically focusing on unserved communities;
- creating incentives for more efficient use of government spectrum and new standards for commercial spectrum to bring affordable broadband to rural communities;
- ensuring that schools, libraries and hospitals have access to next-generation networks and that adequate training and resources are available to enable these institutions to take full advantage of broadband connectivity; and
- encouraging public/private partnerships at the local level that result in broadband to unserved communities [35]

It is likely that these and other potential elements of a national broadband policy, in tandem with broadband investment measures and development of a national broadband strategy by the FCC as directed by the American Recovery and Reinvestment Act of 2009, will significantly shape and expand federal policies and programs intended to promote broadband deployment and adoption.

CURRENT FEDERAL BROADBAND PROGRAMS

Aside from the broadband programs newly established by the American Recovery and Reinvestment Act of 2009 (P.L. 111-5), [36] the Rural Broadband Access Loan and Loan Guarantee Program and the Community Connect Broadband Grants, both at the Rural Utilities Service of the U.S. Department of Agriculture, are currently the only federal programs *exclusively* dedicated to deploying broadband infrastructure. However, there exist other federal programs that provide financial assistance for various aspects of telecommunications development. The major vehicle for funding telecommunications development, particularly in rural and low-income areas, is the Universal Service Fund (USF). While the USF's High Cost Program does not *explicitly* fund broadband infrastructure, subsidies are used, in many cases, to upgrade existing telephone networks so that they are capable of delivering high-speed services. Additionally, subsidies provided by USF's Schools and Libraries Program and Rural Health Care Program are used for a variety of telecommunications services, including broadband access.

Table 1 (at the end of this report) shows selected federal domestic assistance programs throughout the federal government that can be associated with telecommunications development. Many (if not most) of these programs can be related, if not necessarily to the deployment of broadband technologies in particular, then to telecommunications and the "digital divide" issue generally.

Table 2 (also at the end of this report) presents selected federal programs that have provided financial assistance for broadband. These programs are broken down into three categories: first, programs that fund access to telecommunications services in unserved or underserved areas; second, general economic development programs that have funded broadband-related projects; and third, applications-specific programs which will typically fund some aspect of broadband access as a means towards supporting a particular application, such as distance learning or telemedicine.

Rural Utilities Service Program

The Rural Electrification Administration (REA), subsequently renamed the Rural Utilities Service (RUS), was established by the Roosevelt Administration in 1935. Initially, it was established to provide credit assistance for the development of rural electric systems. In 1949, the mission of REA was expanded to include rural telephone providers. Congress further amended the Rural Electrification Act in 1971 to establish within REA a Rural Telephone Account and the Rural Telephone Bank (RTB).

Rural Telephone Loans and Loan Guarantees provide long-term direct and guaranteed loans for telephone lines, facilities, or systems to furnish and improve telecommunications service in rural areas. The RTB—liquidated in FY2006—was a public-private partnership intended to provide additional sources of capital that would supplement loans made directly by RUS. Another program, the Distance Learning and Telemedicine Program, specifically addresses health care and education needs of rural America.

Table 1. Selected Federal Domestic Assistance Programs Related to Telecommunications Development

Program	Agency	Description	FY2008 (obligations)	Web Links for More Information http://12.46.245.173/cfda/cfda.html: Go to "All Programs Listed Numerically" and search by program
Public Telecommunications Facilities—Planning and Construction	National Telecommunications and Information Administration, Dept. of Commerce	Assists in planning, acquisition, installation and modernization of public telecommunications facilities	$19.5 million	http://www.ntia.doc.gov/otiahome/ptfp/index.html
Investments for Public Works and Economic Development Facilities	Economic Development Administration, Dept. of Commerce	Provides grants to economically distressed areas for construction of public facilities and infrastructure, including broadband deployment and other types of telecommunications enabling projects	$249 million	http://www.eda.gov/
Rural Telephone Loans and Loan Guarantees	Rural Utilities Service, U.S. Dept. of Agriculture	Provides long-term direct and guaranteed loans to qualified organizations for the purpose of financing the improvement, expansion, construction, acquisition, and operation of telephone lines, facilities, or systems to furnish and improve telecommunications service in rural areas	$145 million (hardship loans); $250 million (cost of money loans); $295 million (FFB Treasury loans)	http://www.usda.gov/rus/telecom/index.htm

Table 1 (Continued)

Program	Agency Description		FY2008 (obligations) Web Links for More Information http://12.46.245.173/cfda/cfda.html: Go to "All Programs Listed Numerically" and search by program	
Distance Learning and Telemedicine Loans and Grants	Rural Utilities Service, U.S. Dept. of Agriculture	Provides seed money for loans and grants to rural community facilities (e.g., schools, libraries, hospitals) for advanced telecommunications systems that can provide health care and educational benefits to rural areas	$24.7 million (grants) $28 million (loans and loan-grant combinations)	http://www.usda.gov/rus/telecom/dlt/dlt.htm
Rural Broadband Access Loan and Loan Guarantee Program	Rural Utilities Service, U.S. Dept. of Agriculture	Provides loan and loan guarantees for facilities and equipment providing broadband service in rural communities	$300 million (cost of money loans)	http://www.usda.gov/rus/telecom/ broadband.htm
Community Connect Broadband Grants	Rural Utilities Service, U.S. Dept. of Agriculture	Provides grants to applicants proposing to provide broadband service on a "community-oriented connectivity" basis to rural communities of under 20,000 inhabitants.	$13.4 million	http://www.usda.gov/rus/telecom/index.htm
Education Technology State Grants	Office of Elementary and Secondary Education, Dept. of Education	Grants to State Education Agencies for development of information technology to improve teaching and learning in schools	$267 million	http://www.ed.gov/Technology/TLCF/ index.html
Ready to Teach	Office of Assistant Secretary for Educational Research and Improvement, Dept. of Education	Grants to carry out a national telecommunication-based program to improve the teaching in core curriculum areas.	$10.7 million	http://www.ed.gov/programs/readyteach/ index.html

Table 1 (Continued)

Program	Agency	Description	FY2008 (obligations)	Web Links for More Information http://12.46.245.173/cfda/cfda.html: Go to "All Programs Listed Numerically" and search by program
Special Education—Technology and Media Services for Individuals with Disabilities	Office of Special Education and Rehabilitative Services, Dept. of Education	Supports development and application of technology and education media activities for disabled children and adults	$39.3 million	http://www.ed.gov/about/offices/list/osers/index.html?src=mr/
Telehealth Network Grants	Health Resources and Services Administration, Department of Health and Human Services	Grants to develop sustainable telehealth programs and networks in rural and frontier areas, and in medically unserved areas and populations.	$3.9 million	http://www.hrsa.gov/telehealth/
Medical Library Assistance	National Library of Medicine, National Institutes of Health, Department of Health and Human Services	Provides funds to train professional personnel; strengthen library and information services; facilitate access to and delivery of health science information; plan and develop advanced information networks; support certain kinds of biomedical publications; and conduct research in medical informatics and related sciences	$67.5 million	http://www.nlm.nih.gov/ep/extramural.html

Table 1 (Continued)

Program	Agency	Description	FY2008 (obligations)	Web Links for More Information http://12.46.245.173/cfda/cfda.html: Go to "All Programs Listed Numerically" and search by program
State Library Program	Office of Library Services, Institute of Museum and Library Services, National Foundation on the Arts and the Humanities	Grants to state library administrative agencies for promotion of library services that provide all users access to information through State, regional, and international electronic networks	$171.5 million	http://www.imls.gov/grants/library/ lib_gsla.asp#po
Program	Agency	Description	FY2008 (obligations)	Web Links for More Information http://12.46.245.173/cfda/cfda.html: Go to "All Programs Listed Numerically" and search by program
Native American and Native Hawaiian Library Services	Office of Library Services, Institute of Museum and Library Services, National Foundation on the Arts and the Humanities	Supports library services including electronically linking libraries to networks	$3.7 million	http://www.imls.gov/grants/library/lib_nat.asp
Appalachian Area Development	Appalachian Regional Commission	Provides project grants for Appalachian communities to support the physical infrastructure necessary for economic development and improved quality of life.	$73 million	http://www.arc.gov/index.do?nodeId=21

Table 1 (Continued)

Program	Agency	Description	FY2008 (obligations)	Web Links for More Information http://12.46.245.173/cfda/cfda.html: Go to "All Programs Listed Numerically" and search by program
Delta Area Economic Development	Delta Regional Authority	Grants to support self-sustaining economic development of eight states in Mississippi Delta region.	$7.8 million	http://www.dra.gov/programs/informationtechnology
Denali Commission Program	Denali Commission	Provides grants through a federal and state partnership designed to provide critical infrastructure and utilities throughout Alaska, particularly in distressed communities	$106 million	http://www.denali.gov/

Source: Prepared by CRS based on information from the Catalog of Federal Domestic Assistance.

Table 2. Selected Federal Programs Funding Broadband Access

Program	Comments
Programs Funding Access to Telecommunications in Underserved Areas	
Rural Broadband Access Loan and Loan Guarantee Program (Rural Utilities Service, U.S. Department of Agriculture) Community Connect Broadband Grants (Rural Utilities Service, U.S. Department of Agriculture) Rural Telephone Loans and Loan Guarantees (Rural Utilities Service, U.S. Department of Agriculture) Universal Service Fund: High Cost Program (Federal Communications Commission)	Provides loan and loan guarantees for facilities and equipment providing broadband service in rural communities. Provides grants to applicants proposing to provide broadband service on a "community-oriented connectivity" basis to rural communities of under 20,000 inhabitants. Since 1995, the RUS Rural Telephone Loan and Loan Guarantee program— which has traditionally financed telephone voice service in rural areas under 5,000 inhabitants—has required that all telephone facilities receiving financing must be capable of providing DSL broadband service at a rate of at least 1 megabyte per second. While the USF's High Cost Program does not explicitly fund broadband infrastructure, subsidies are used, in many cases, to upgrade existing telephone networks.
Federal Economic Development Programs Funding Broadband Access	
Community Development Block Grants (Department of Housing and Urban Development) Indian Community Development Block Grants (Department of Housing and Urban Development) Grants for Public Works and Economic Development Facilities (Economic Development Administration, Department of Commerce) Appalachian Regional Commission	In Michigan, a Digital Divide Investment Program (DDIP) combined Michigan Broadband Development Authority loans (initially $12 million) and CDBG grant funding ($4 million) to deploy a hybrid fixed wireless and fiber network in two rural counties which would make broadband affordable for low to moderate income residents. In 2005, HUD awarded the Coquille Indian Tribe a $421,354 grant used to fund the Coquille Broadband Technology Infrastructure Project. The project will allow for improved connectivity for reservation residents, improvements in rural community access, and potentially increased wireless Internet access for the Tribal and surrounding communities. Supports the proliferation of broadband networks as a key priority for regional economic growth. Examples: $6 million grant to a company in Virginia for investment in 300 miles of fiber optic cable in nine counties and three cities; $2 million grant to companies in Vermont to help build a 424 mile fiber optic broadband network in rural northern Vermont; and $270 thousand to support a Rhode Island Wireless Innovation Networks project. EDA encourages communities eligible for RUS programs to access that first before applying for EDA investment dollars. The Appalachian

Table 2 (Continued)

Program Comments	
	Regional Development Act Amendments of 2002 reauthorized ARC for five years and created specific authority for a Region-wide initiative to bridge the telecommunications and technology gap between the Appalachian Region and the rest of the United States. Supported a telecommunications initiative ($33 million over five year period) which includes projects such as: a regional fiber network across northeast Mississippi; wireless demonstrations in rural New York, Ohio, Pennsylvania, Virginia, West Virginia, and Georgia; and a regionwide effort in Kentucky to compile an inventory of broadband access across the 51 Appalachian counties and work with the private sector to substantially increase broadband coverage. In Maryland, a county-wide high-speed wireless network, funded by ARC over several years, now serves over 4,500 customers
Delta Regional Authority Denali Commission	During a strategic planning retreat in February 2005, the DRA board determined that one of the authority's three top policy priorities would be information technology. To support its policy position, the authority devoted $150,000 to create an information technology plan for the region. Funded Telecommunications Survey in 2000 which was used to determine the state of broadband deployment in Alaska and used as basis for applying for RUS broadband assistance.
Applications-Based Federal Programs Related to Broadband	
Universal Service Fund: Schools and Libraries or "E-Rate" Program (Federal Communications Commission) Universal Service Fund: Rural Health Care Program (Federal Communications Commission) Distance Learning and Telemedicine Program (Rural Utilities Service, U.S. Department of Agriculture) Public Safety Interoperable Communications Grant Program (National Telecommunications and Information Administration, Department of Commerce)	Used to fund broadband access for schools and libraries. Used to fund broadband access for rural health care centers. Provides seed money for loans and grants to rural community facilities (e.g., schools, libraries, hospitals) for advanced telecommunications systems that can provide health care and educational benefits to rural areas. Provides funding to states and territories to enable and enhance public safety agencies' interoperable communications capabilities. Grants to develop sustainable telehealth programs and networks in rural and frontier areas, and in medically unserved areas and populations. Grants for public television, public radio, and nonbroadcast distance learning projects. Examples include Education Technology State Grants, Ready to Teach. Grants to state library administrative agencies for promotion of library services

Table 2 (Continued)

Program Comments	
Telehealth Network Grants (Health Resources and Services Administration, Department of Health and Human Services) Public Telecommunications Facilities Program (National Telecommunications and Information Administration, Department of Commerce) Education technology programs (Department of Education) State Library Grants (Office of Library Services, Institute of Museum and Library Services, National Foundation on the Arts and the Humanities) Medical Library Assistance (National Library of Medicine, National Institutes of Health, Department of Health and Human Services)	that provide all users access to information through State, regional, and international electronic networks. Provides funds to train professional personnel; strengthen library and information services; facilitate access to and delivery of health science information; plan and develop advanced information networks; support certain kinds of biomedical publications; and conduct research in medical informatics and related sciences.

RUS implements two programs specifically targeted at providing assistance for broadband deployment in rural areas: the Rural Broadband Access Loan and Loan Guarantee Program and Community Connect Broadband Grants. The 110th Congress reauthorized and reformed the Rural Broadband Access Loan and Loan Guarantee program as part of the 2008 farm bill (P.L. 110-234). For further information on rural broadband and the RUS broadband programs, see CRS Report RL33816, *Broadband Loan and Grant Programs in the USDA's Rural Utilities Service*, by Lennard G. Kruger.

The Universal Service Concept and the FCC [37]

Since its creation in 1934 the Federal Communications Commission (FCC) has been tasked with "... mak[ing] available, so far as possible, to all the people of the United States, ... a rapid, efficient, Nation-wide, and world-wide wire and radio communications service with adequate facilities at reasonable charges.... " [38] This mandate led to the development of what has come to be known as the universal service concept.

The universal service concept, as originally designed, called for the establishment of policies to ensure that telecommunications services are available to all Americans, including those in rural, insular and high cost areas, by ensuring that rates remain affordable. Over the years this concept fostered the development of various FCC policies and programs to meet this goal. The FCC offers universal service support through a number of direct mechanisms that target both providers of and subscribers to telecommunications services [39].

The development of the federal universal service high cost fund is an example of provider-targeted support. Under the high cost fund, eligible telecommunications carriers, usually those serving rural, insular and high cost areas, are able to obtain funds to help offset the higher than average costs of providing telephone service [40].

This mechanism has been particularly important to rural America where the lack of subscriber density leads to significant costs. FCC universal service policies have also been expanded to target individual users. Such federal programs include two income-based programs, Link Up and Lifeline, established in the mid-1980s to assist economically needy individuals. The Link Up program assists low-income subscribers pay the costs associated with the initiation of telephone service and the Lifeline program assists low-income subscribers pay the recurring monthly service charges.

Funding to assist carriers providing service to individuals with speech and/or hearing disabilities is also provided through the Telecommunications Relay Service Fund. Effective January 1, 1998, schools, libraries, and rural health care providers also qualified for universal service support.

The Universal Service and the Telecommunications Act of 1996

Passage of the Telecommunications Act of 1996 (P.L. 104-104) codified the long-standing commitment by U.S. policymakers to ensure universal service in the provision of telecommunications services.

The Schools and Libraries, and the Rural Health Care Programs

Congress, through the 1996 Act, not only codified, but also expanded the concept of universal service to include, among other principles, that elementary and secondary schools and classrooms, libraries, and rural health care providers have access to telecommunications services for specific purposes at discounted rates. (See Sections 254(b)(6) and 254(h)of the 1996 Telecommunications Act, 47 U.S.C. 254.)

1. The Schools and Libraries Program. Under universal service provisions contained in the 1996 Act, elementary and secondary schools and classrooms and libraries are designated as beneficiaries of universal service discounts. Universal service principles detailed in Section 254(b)(6) state that "Elementary and secondary schools and classrooms ... and libraries should have access to advanced telecommunications services...." The act further requires in Section 254(h)(1)(B) that services within the definition of universal service be provided to elementary and secondary schools and libraries for education purposes at discounts, that is at "rates less than the amounts charged for similar services to other parties."

The FCC established the Schools and Libraries Division within the Universal Service Administrative Company (USAC) to administer the schools and libraries or "E (education)-rate" program to comply with these provisions. Under this program, eligible schools and libraries receive discounts ranging from 20 to 90 percent for telecommunications services depending on the poverty level of the school's (or school district's) population and its location in a high cost telecommunications area. Three categories of services are eligible for discounts: internal connections (e.g., wiring, routers and servers); Internet access; and telecommunications and dedicated services, with the third category receiving funding priority. According to data released by program administrators, $21.3 billion in funding has been committed over the first ten years of the program with funding released to all states, the District of Columbia and all territories. Funding commitments for funding Year 2008 (July 1, 2008 - June 30, 2009), the eleventh and current year of the program, totaled $2.2 billion as of March 10, 2009 [41].

2. The Rural Health Care Program. Section 254(h) of the 1996 Act requires that public and non-profit rural health care providers have access to telecommunications services necessary for the provision of health care services at rates comparable to those paid for similar services in urban areas. Subsection 254(h)(1) further specifies that "to the extent technically feasible and economically reasonable" health care providers should have access to advanced telecommunications and information services. The FCC established the Rural Health Care Division (RHCD) within the USAC to administer the universal support program to comply with these provisions. Under FCC established rules only public or non-profit health care providers are eligible to receive funding. Eligible health care providers, with the exception of those requesting only access to the Internet, must also be located in a rural area. The funding ceiling, or cap, for this support was established at $400 million annually. The funding level for Year One of the program (January 1998 - June 30, 1999) was set at $100 million. Due to less than anticipated demand, the FCC established a $12 million funding level for the second year (July 1, 1999 to June 30, 2000) of the program but has since returned to a $400 million yearly cap. As of March 17, 2009, covering the first 11 years of the program, a total of $284.7 million has been committed to 4,167 rural health care providers. The primary use of the funding is to provide reduced rates for telecommunications and information services necessary for the provision of health care [42].

The Telecommunications Development Fund

Section 714 of the 1996 Act created the Telecommunications Development Fund (TDF). The TDF is a private, non-governmental, venture capital corporation currently overseen by a five-member board of directors and fund management. The TDF focuses on seed, early stage, and select later stage investments in communications and has $90 million under management in two funds. Fund I is no longer making new investments. Fund II remains active and currently has 13 companies in its investment portfolio Funding is largely derived from the interest earned from the upfront payments bidders submit to participate in FCC auctions. The TDF also provides entrepreneur education, training, management and technical assistance in underserved rural and urban communities through the TDF Foundation [43].

Universal Service and Broadband

One of the policy debates surrounding universal service is whether access to advanced telecommunications services (i.e. broadband) should be incorporated into universal service objectives. The term universal service, when applied to telecommunications, refers to the ability to make available a basket of telecommunications services to the public, across the nation, at a reasonable price. As directed in the 1996 Telecommunications Act [Section 254(c)] a federal-state Joint Board was tasked with defining the services which should be included in the basket of services to be eligible for federal universal service support; in effect using and defining the term "universal service" for the first time. The Joint Board's recommendation, which was subsequently adopted by the FCC in May 1997, included the following in its universal service package: voice grade access to and some usage of the public switched network; single line service; dual tone signaling; access to directory assistance; emergency service such as 911; operator services; access and interexchange (long distance) service.

Some policy makers expressed concern that the FCC-adopted definition is too limited and does not take into consideration the importance and growing acceptance of advanced services such as broadband and Internet access. They point to a number of provisions contained in the Universal Service section of the 1996 Act to support their claim. Universal service principles contained in Section 254(b)(2) state that "Access to advanced telecommunications services should be provided to all regions of the Nation." The subsequent principle (b)(3) calls for consumers in all regions of the nation including "low-income" and those in "rural, insular, and high cost areas" to have access to telecommunications and information services including "advanced services" at a comparable level and a comparable rate charged for similar services in urban areas. Such provisions, they state, dictate that the FCC expand its universal service definition.

Others caution that a more modest approach is appropriate given the "universal mandate" associated with this definition and the uncertainty and costs associated with mandating nationwide deployment of such advanced services as a universal service policy goal. Furthermore they state the 1996 Act does take into consideration the changing nature of the telecommunications sector and allows for the universal service definition to be modified if future conditions warrant. Section 254(c)of the act states that "universal service is an evolving level of telecommunications services" and the FCC is tasked with "periodically" reevaluating this definition "taking into account advances in telecommunications and information technologies and services." Furthermore, the Joint Board is given specific authority to recommend "from time to time" to the FCC modification in the definition of the services to be included for federal universal service support. The Joint Board, on November 19, 2007, concluded such an inquiry and recommended that the FCC change the mix of services eligible for universal service support.

The Joint Board recommended, among other things, that "the universal availability of broadband Internet services" be included in the nation's communications goals and hence be supported by federal universal service funds [44]. In response to the Joint Board recommendation, the FCC, on January 29, 2008, released three notices of proposed rulemaking dealing with specific aspects of universal service, including an examination of the scope of the definition.

The FCC is still examining proposals for universal service reform, including expanding the program to include broadband, but has not taken action.

LEGISLATION IN THE 100TH CONGRESS

In the 110th Congress, legislation was introduced that would provide financial assistance for broadband deployment. Of particular note is the reauthorization of the Rural Utilities Service (RUS) broadband loan program, which was enacted as part of the 2008 farm bill (P.L. 110-234). In addition to reauthorizing and reforming the RUS broadband loan program, P.L. 110-234 contains provisions establishing a National Center for Rural Telecommunications Assessment and requiring the FCC and RUS to formulate a comprehensive rural broadband strategy.

The Broadband Data Improvement Act (P.L. 110-385) was enacted by the 110th Congress and requires the FCC to collect demographic information on unserved areas, data comparing broadband service with 75 communities in at least 25 nations abroad, and data on consumer use of broadband. The act also directs the Census Bureau to collect broadband data, the Government Accountability Office to study broadband data metrics and standards, and the

Department of Commerce to provide grants supporting state broadband initiatives.

Meanwhile, the America COMPETES Act (H.R. 2272) was enacted (P.L. 110-69) and contains a provision authorizing the National Science Foundation (NSF) to provide grants for basic research in advanced information and communications technologies. Areas of research include affordable broadband access, including wireless technologies. P.L. 110-69 also directs NSF to develop a plan that describes the current status of broadband access for scientific research purposes.

The following is a listing of broadband related bills enacted in the 110th Congress.

P.L. 110-69 (H.R. 2272)

America COMPETES Act. Authorizes the National Science Foundation (NSF) to provide grants for basic research in advanced information and communications technologies. Areas of research include affordable broadband access, including wireless technologies. Also directs NSF to develop a plan that describes the current status of broadband access for scientific research purposes. Introduced May 10, 2007; referred to House Committee on Science and Technology. Passed House May 21, 2007. Passed Senate July 19, 2007. Signed into law, August 9, 2007.

P.L. 110-161 (H.R. 2764)

Consolidated Appropriations Act, 2008. For Rural Utilities Service, U.S. Department of Agriculture, provides $6.45 million to support a loan level of $300 million for the broadband loan program, and $13.5 million for broadband community connect grants. Signed by President, December 26, 2007.

P.L. 110-234 (H.R. 2419)

Food, Conservation, and Energy Act of 2008. Reauthorizes broadband program at the Rural Utilities Service through FY2012. Establishes a National Center for Rural Telecommunications Assessment. Directs USDA and the FCC to submit to Congress a comprehensive rural broadband strategy.

Introduced May 22, 2007; referred to Committee on Agriculture, and in addition to Committee on Foreign Affairs. Subcommittee on Specialty Crops, Rural Development, and Foreign Agriculture held markup of Title VII (Rural Development) on June 6, 2007. Reported by House Committee on Agriculture (H.Rept. 110-256) on July 23, 2007. Passed House July 27, 2007. Passed Senate with an amendment, December 14, 2007. Conference report (H.Rept. 110-627) approved by the House May 14, 2008, and by the Senate May 15, 2008. Vetoed by the President, May 21, 2008. House and Senate overrode veto on May 21 and May 22, 2008. Became P.L. 110-234, May 22, 2007.

P.L. 110-329 (H.R. 2638)

Consolidated Security, Disaster Assistance, and Continuing Appropriations Act, 2009. Continuing resolution funds RUS broadband loan and grant program at FY2008 levels through March 6, 2009. Signed by President September 30, 2008.

P.L. 110-385 (S. 1492)

Broadband Data Improvement Act. Seeks to improve the quality of federal broadband data collection and encourage state initiatives that promote broadband deployment. Requires the FCC to collect demographic information on unserved areas, data comparing broadband service with 75 communities in at least 25 nations abroad, and data on consumer use of broadband. Directs the Census Bureau to collect broadband data, the Government Accountability Office to study broadband data metrics and standards, and the Department of Commerce to provide grants supporting state broadband initiatives. Introduced May 24, 2007; referred to Committee on Commerce, Science, and Transportation. Ordered to be reported July 19, 2007; reported by Committee (S.Rept. 110-204) and placed on Senate Legislative Calendar, October 24, 2007. Passed by Senate with an amendment September 26, 2008. Passed by House September 29, 2008. Became P.L. 110-385, October 10, 2008.

BROADBAND STIMULUS LEGISLATION IN THE 111TH CONGRESS

In December 2008, leadership in the House and Senate, as well as the Obama transition team, announced their intention to include a broadband component in the infrastructure portion of the economic stimulus package. At the same time, numerous interested parties, including: broadband equipment manufacturers; large, mid-sized, and small wireline and wireless service providers; satellite operators; telecommunications unions; consumer groups; education groups; public safety organizations; think tanks; and others unveiled a multitude of specific proposals for government support of broadband infrastructure [45].

House

On January 21, 2009, the House Appropriations Committee approved legislative language for the spending portion of the economic stimulus package (H.R. 1, American Recovery and Reinvestment Act of 2009). The legislation would provide $6 billion to support deployment of broadband and wireless services in rural, unserved, and underserved areas of the nation. Of the total, $2.825 billion would be provided to the Rural Utilities Service (RUS) of the Department of Agriculture, and $3.175 billion to the National Telecommunications and Information Administration of the Department of Commerce. The House broadband stimulus provisions are included within Title II (under Rural Utilities Service), Title III (under National Telecommunications and Information Administration), and Title VI (Broadband Communications) of H.R. 1. Specifically, the legislation breaks down as follows:

- *$2.825 billion* to the Rural Utilities Service for additional loans, loan guarantees, and grants to finance "open access" broadband infrastructure. Specifies that at least 75% of the area to be served by a project receiving funds shall be in a rural area without sufficient access to high speed broadband service to facilitate economic development, as determined by the Secretary of Agriculture. Priority is given to projects that provide service to the most rural residents that do not have access to broadband services. Priority is also given to borrowers and former borrowers of rural telephone loans.

- *$350 million* to the National Telecommunications and Information Administration to establish the State Broadband Data and Development Grant Program, as authorized by the recently enacted Broadband Data Improvement Act (P.L. 110-385). Grants would be used to develop and implement statewide initiatives to identify and track the availability and adoption of broadband within each state. Would also be used to develop and maintain a nationwide broadband inventory map.
- *$1 billion* to NTIA for Wireless Deployment Grants for wireless voice service and advanced wireless broadband service (at least 3 Mbps downstream, 1 Mbps upstream). To the extent possible, 25% of the grants are to be awarded for providing wireless voice service in unserved areas, and 75% for advanced wireless broadband service in underserved areas. Grant recipients are required to operate on an "wireless open access" basis.
- *$1.825 billion* to NTIA for Broadband Deployment Grants for basic broadband service (at least 5 Mbps downstream, 1 Mbps upstream) or advanced broadband service (at least 45 Mbps downstream, 15 Mbps upstream). To the extent possible, 25% of the grants are to be awarded for providing basic broadband in unserved areas, and 75% for advanced broadband service in underserved areas. Grant recipients are required to operate on an "open access" basis.

For the Wireless and Broadband Deployment Grants, the terms "unserved," "underserved," "open access," and "wireless open access" shall be defined by the FCC not later than 45 days after enactment of the legislation. Also for these grants, each state planning to participate is required to submit to NTIA a report detailing which geographic areas within that state are most in need of wireless voice, advanced wireless broadband, basic broadband, and advanced broadband services in both unserved and underserved areas. Unserved and underserved areas identified by a state shall not constitute more than 20% of the population or geographic area of that state.

While the RUS broadband programs and the State Broadband Data and Development Grant Program were previously authorized (the RUS programs have operated for seven years, while the state grants is newly established by P.L. 110-385, the Broadband Data Improvement Act, and not yet funded), the

Broadband Deployment Grants and the Wireless Deployment Grants would be newly authorized.

On January 22, 2009, the House Committee on Energy and Commerce marked up and approved sections 3101 (nationwide broadband inventory map to be developed by NTIA) and 3102 (authorizing wireless and broadband deployment grants at NTIA). An amendment in the nature of a substitute, offered by the Chairman, additionally requires NTIA to issue an annual report assessing the impact and effectiveness of the grants, and expands the list of eligible entities to include satellite companies. Other amendments agreed to by the Committee would:

- include the improvement of interoperable broadband communications systems used for public safety and emergency response among factors to be considered in award decisions;
- direct the FCC to review and revise its definitions of unserved and underserved areas after completion of NTIA's nationwide broadband inventory map;
- direct the FCC to submit to Congress a national broadband plan; and
- direct NTIA to consider whether an eligible entity is a socially and economically disadvantaged small business.

On January 28, 2009, the House passed H.R. 1. An amendment agreed to by the House would make projects funded by the newly established NTIA broadband and wireless grant programs eligible for worker training grant money (under Title IX, Subtitle A of H.R. 1).

Senate

On February 7, 2009, a substitute amendment to H.R. 1 (S.Amdt. 570, the "Collins-Nelson amendment") was submitted in the Senate. S.Amdt. 570 would provide $7 billion to NTIA for establishment of a national broadband service development and expansion program called the Broadband Technology Opportunities Program. This is $2 billion less than what was provided in the Senate Appropriations Committee mark (S. 336, S.Rept. 111-3). The program, as provided in S.Amdt. 570, consists of:

- *$6.650 billion* for competitive broadband grants, of which not less than $200 million shall be available for competitive grants for expanding public computer center capacity (including at community colleges and public libraries); not less than $250 million to encourage sustainable adoption of broadband service; and $10 million transferred to the Department of Commerce Office of Inspector General for audits and oversight.

- *$350 million* for funding the Broadband Data Improvement Act (P.L. 110-385) and for the purpose of developing and maintaining a broadband inventory map, which shall be made accessible to the public no later than two years after enactment. Funds deemed necessary and appropriate by the Secretary of Commerce may be transferred to the FCC for the purposes of developing a national broadband plan, which shall be completed one year after enactment.

Significant language related to Broadband Technology Opportunities Program grants includes the following:

- 50% of the total grant funding shall be used to support projects in rural communities, and funds may be transferred for this purpose to USDA's Rural Utilities Service if deemed appropriate by the Secretary of Commerce and in consultation with the Secretary of Agriculture. In cases where this funding is made available to the RUS broadband loan, loan guarantee, and grant programs, at least 75% of the area to be served by a project receiving funds shall be in a rural area without sufficient access to high speed broadband service to facilitate economic development, as determined by the Secretary of Agriculture. Priority is given to projects that provide service to the highest proportion of rural residents that do not have sufficient access to broadband services. Among the purposes of the grant program is to provide broadband to citizens residing in unserved and underserved areas. NTIA may consult with the chief executive officer of any state with respect to identifying unserved and underserved areas within that state, and with respect to the allocation of grant funds within that state.
- NTIA shall, in coordination with the FCC, develop nondiscrimination and network interconnection obligations that shall be contractual conditions of grants awarded.
- The federal share of any project may not exceed 80% unless NTIA determines financial hardship.

- Grant eligibility includes: a state or political subdivision, a nonprofit foundation, corporation, institution or association, Indian tribe, Native Hawaiian organization or other nongovernmental entity in partnership with a state or political subdivision, Indian tribe, or native Hawaiian organization.
- Grants may be used to acquire equipment and technology necessary for broadband infrastructure, to construct and deploy broadband service related infrastructure, to ensure access to broadband by community anchor institutions, to facilitate broadband service by vulnerable populations, to construct and deploy broadband facilities to improve public safety communications.

S.Amdt. 570 also includes an investment tax credit for qualified broadband expenditures. The provision would establish a 10% tax credit for investment in current generation broadband in rural and underserved areas, a 20% tax credit for investment in current generation broadband in unserved areas, and a 20% tax credit for investment in next generation broadband in rural, underserved, and unserved areas. Current generation is defined as at least 5 Mbps download speed and 1 Mbps upload, or for wireless broadband, 3Mbps download and 768 kbps upload. Next generation is defined as at least 100 Mbps download and 20 Mbps upload.

On February 10, 2009, the Senate passed H.R. 1 as amended by S.Amdt. 570.

Comparison of House and Senate Bills

The following are some key similarities and differences between the House-passed and Senate-passed broadband provisions of H.R. 1:

- both the House and Senate bills would rely primarily on grants as a strategy to stimulate broadband deployment – House total funding is $6 billion, Senate total is $7 billion;
- House would provide $3.175 billion to NTIA and $2.825 billion to RUS broadband programs; Senate provides all funding to NTIA, but directs that 50% should finance projects in rural areas, DOC can transfer this funding in part to the RUS broadband loan and grant programs if deemed appropriate;

- both the House and Senate bills would provide $350 million to NTIA for funding the Broadband Data Improvement Act and to develop a national broadband inventory map;
- Senate specifically sets aside not less than $200 million for competitive grants for expanding public computer center capacity (including at community colleges and public libraries) and not less than $250 million to encourage sustainable adoption of broadband service; funding is not specifically set aside for these purposes in the House bill;
- Senate has a 20% matching requirement for grant recipients (which can be waived in case of financial hardship); House doesn't have a matching requirement but cites a 20% match as a positive consideration when assessing grant applications;
- House sets funding allocation percentages for Broadband Technology Opportunity grants based on minimum broadband speed requirements (download and upload) and whether area is unserved or underserved, directs FCC to define "unserved" and "underserved" within 45 days; Senate doesn't prescribe allocations based on minimum download/ upload speeds and whether an area is unserved or underserved, instead directs NTIA to consult with each state to identify unserved and underserved areas as well as the appropriate allocation of grant funds within that state;
- House mandates "open access" requirement for grant projects and requires that projects adhere to FCC's net neutrality principles, directs FCC to define "open access" and "wireless open access" within 45 days; Senate directs that NTIA shall, in coordination with the FCC, develop nondiscrimination and network interconnection obligations that shall be contractual conditions of grants awarded;
- House defines entities eligible for grants as essentially any provider of wireless or broadband service, including states or local governments; Senate defines an eligible applicant as a state or political subdivision thereof, a nonprofit foundation, corporation, institution or association, Indian tribe, Native Hawaiian organization or other nongovernmental entity in partnership with a state or political subdivision, Indian tribe, or native Hawaiian organization;
- Senate includes broadband investment tax credits; House does not include broadband tax incentives;

- House directs that 50% of grant funds are to be awarded no later than September 30, 2009; Senate directs all funds to be awarded by the end of FY2010; and
- both the House and Senate bills direct FCC to develop a national broadband plan in one year.

P.L. 111-5: The American Recovery and the Reinvestment Act of 2009

On February 17, 2009, President Obama signed P.L. 111-5, the American Recovery and Reinvestment Act (ARRA). Broadband provisions of the ARRA provide a total of $7.2 billion, primarily for broadband grants. The total consists of $4.7 billion to NTIA/DOC for a newly established Broadband Technology Opportunities Program and $2.5 billion to existing RUS/USDA broadband programs [46].

Regarding the $2.5 billion to RUS/USDA broadband programs, the ARRA specifies that at least 75% of the area to be served by a project receiving funds shall be in a rural area without sufficient access to high speed broadband service to facilitate economic development, as determined by the Secretary of Agriculture. Priority is given to projects that provide service to the most rural residents that do not have access to broadband services. Priority is also given to borrowers and former borrowers of rural telephone loans.

Of the $4.7 billion appropriated to NTIA:

- $4.35 billion is directed to a competitive broadband grant program, of which not less than $200 million shall be available for competitive grants for expanding public computer center capacity (including at community colleges and public libraries); not less than $250 million to encourage sustainable adoption of broadband service; and $10 million transferred to the Department of Commerce Office of Inspector General for audits and oversight; and
- $350 million is directed for funding the Broadband Data Improvement Act (P.L. 110-385) and for the purpose of developing and maintaining a broadband inventory map, which shall be made accessible to the public no later than two years after enactment. Funds deemed necessary and appropriate by the Secretary of Commerce may be transferred to the FCC for the purposes of developing a national broadband plan, which shall be completed one year after enactment.

The Broadband Technology Opportunities Program within NTIA is authorized by Division B, Title VI of the ARRA. Specific implementation requirements and guidelines for the new NTIA broadband grants are as follows:

- Directs NTIA to consult with each state to identify unserved and underserved areas (with respect to access to broadband service) as well as the appropriate allocation of grant funds within that state. The Conferees (H.Rept. 111-16) "intend that the NTIA has discretion in selecting the grant recipients that will best achieve the broad objectives of the program."
- The substitute does not define "unserved area," "underserved area," and "broadband." The Conferees instructed NTIA to coordinate its understanding of these terms with the FCC, and in defining "broadband service" to take into consideration technical differences between wireless and wireline networks and to consider the actual speeds these networks are able to deliver to consumers under a variety of circumstances.
- Directs NTIA, in coordination with the FCC, to publish "non-discrimination and network interconnection obligations" that shall be contractual conditions of awarded grants, and specifies that these obligations should adhere, at a minimum, to the FCC's broadband principles to promote the openness and interconnected nature of the Internet (FCC 05-151, adopted August 5, 2005).
- Directs NTIA, when considering applications for grants, to consider whether the project will provide the greatest broadband speed possible to the greatest population of users in the area. There are no specific speed thresholds that applicants must meet to be eligible for a grant. The Conferees acknowledged that while speed thresholds could have the unintended effect of thwarting broadband deployment in some areas, deploying next-generation speeds would likely result in greater job creation and job preservation. NTIA is instructed to "seek to fund, to the extent practicable, projects that provide the highest possible, next-generation broadband speeds to consumers."
- Defines entities eligible for grants as: a state or political division thereof; the District of Columbia; a territory or possession of the United States; an Indian tribe or native Hawaiian organization; a nonprofit foundation, corporation, institution or association; or any other entity, including a broadband service or infrastructure provider,

that NTIA finds by rule to be in the public interest. It was the intent of the Conferees that as many entities as possible be eligible to apply for a grant, including wireless carriers, wireline carriers, backhaul providers, satellite carriers, public-private partnerships, and tower companies.

- Requires NTIA to consider whether a grant applicant is a socially and economically disadvantaged small business as defined under the Small Business Act.
- Directs NTIA to ensure that all awards are made before the end of FY2010. Grantees will be required to substantially complete projects within two years after the grant is awarded.
- Directs that the federal share of any project cannot exceed 80% unless the applicant petitions NTIA and demonstrates financial need.

The Conference Agreement and public law bill did not include the broadband investment tax credit provisions that were contained in the Senate bill. For more information on implementation of the broadband provisions of the ARRA, see CRS Report R40436, *Broadband Infrastructure Programs in the American Recovery and Reinvestment Act*, by Lennard G. Kruger.

OTHER LEGISLATION IN THE 111TH CONGRESS

P.L. 111-8 (H.R. 1105). Omnibus Appropriations Act, 2009. Appropriates to RUS/USDA $15.619 million to support a loan level of $400.487 million for the Rural Broadband Access Loan and Loan Guarantee Program, and $13.406 million for the Community Connect Grant Program. To the FCC, designates not less than $3 million to establish and administer a State Broadband Data and Development matching grants program for State-level broadband demand aggregation activities and creation of geographic inventory maps of broadband service to identify gaps in service and provide a baseline assessment of statewide broadband deployment. Passed House February 25, 2009. Passed Senate March 10, 2009. Signed by President, March 12, 2009.

H.R. 691 (Meeks). Broadband Access Equality Act of 2009. Amends the Internal Revenue Code of 1986 to provide credit against income tax for businesses furnishing broadband services to underserved and rural areas. Introduced January 26, 2009; referred to Committee on Ways and Means.

H.R. 760 (Eshoo). Advanced Broadband Infrastructure Bond Initiative of 2009. Amends the Internal Revenue Code of 1986 to provide an income tax credit to holders of bonds financing new advanced broadband infrastructure. Introduced January 28, 2009; referred to Committee on Ways and Means and in addition to Committee on Energy and Commerce.

CONCLUDING OBSERVATIONS

As Congress considers various options for encouraging broadband deployment, a key issue is how to strike a balance between providing federal assistance for unserved and underserved areas where the private sector may not be providing acceptable levels of broadband service, while at the same time minimizing any deleterious effects that government intervention in the marketplace may have on competition and private sector investment. In addition to loans, loan guarantees, and grants for broadband infrastructure deployment, a wide array of policy instruments are available to policymakers including universal service reform, tax incentives to encourage private sector deployment, broadband bonds, demand-side incentives (such as assistance to low income families for purchasing computers), regulatory and deregulatory measures, and spectrum policy to spur roll-out of wireless broadband services. In assessing federal incentives for broadband deployment, Congress will likely consider the appropriate mix of broadband deployment incentives to create jobs in the short and long term, the extent to which incentives should target next-generation broadband technologies, the extent to which "underserved" areas with existing broadband providers should receive federal assistance, and how broadband stimulus measures of the ARRA might fit into the context of overall goals for a national broadband policy.

REFERENCES

[1] The term "digital divide" can also refer to international disparities in access to information technology. This report focuses on domestic issues only.
[2] FCC, *High-Speed Services for Internet Access: Status as of December 31, 2007,* January 2009. Available athttp://hraunfoss.fcc.gov/ edoc s_public/attachmatch/DOC-287962A1.pdf.
[3] Initially, and for many years following, the FCC defined broadband (or more specifically "high-speed lines") as over200 kilobits per second (kbps) in at least one direction, which was roughly four times the speed

of conventional dialup Internet access. In recent years, the 200 kbps threshold was considered too low, and on March 19, 2008, the FCC adopted a report and order (FCC 08-89) establishing new categories of broadband speed tiers for data collection purposes. Specifically, 200 kbps to 768 kbps will be considered "first generation," 768 kbps to 1.5 Mbps as "basicbroadband tier 1," and increasingly higher speed tiers as broadband tiers 2 through 7 (tier seven is greater than or equalto 100 Mbps in any one direction). Tiers can change as technology advances.

[4] Horrigan, John, Pew Internet and American Life Project, "Barriers to Broadband Adoption—The User Perspective,"December 19, 2008, available at http://otrans.3cdn.net/fe2b6b302960dbe0d7_bqm6ib242.pdf.

[5] S.Derek Turner, Free Press, *Down Payment on Our Digital Future*, December 2008, p.8.

[6] Federal Communications Commission, *Fourth Report to Congress*, "Availability of Advanced Telecommunications Capability in the United States," GN Docket No. 04-54, FCC 04-208, September 9, 2004, p. 38. Available athttp://hraunfoss.fcc.gov/edocs_public/attachmatch/FCC-04-208A1.pdf.

[7] "Barriers to Broadband Adoption—The User Perspective," p. 1.

[8] Horrigan, John, Pew Internet and American Life Project, "Obama's Online Opportunities II: If You Build It Will They Log On?" January 21, 2009, available at http://www.pewinternet.org/pdfs/PIP_ Broadb and%20Barriers.pdf.

[9] For more information on rural broadband and broadband programs at the Rural Utilities Service, see CRS Report RL33816, *Broadband Loan and Grant Programs in the USDA's Rural Utilities Service*, by Lennard G. Kruger.

[10] See for example: National Exchange Carrier Association (NECA), Trends 2006: Making Progress With Broadband, 2006, 26 p. Available at http://www.neca.org/media/trends_brochure_website.pdf.

[11] Horrigan, John B., Pew Internet and American Life Project, *Home Broadband Adoption 2008,* July 2008, p. 3.Available at http://w ww.pewinternet.org/pdfs/PIP_Broadband_2008.pdf.

[12] FCC, *High-Speed Services for Internet Access: Status as of December 31, 2007,* p. 4.

[13] The terrain of rural areas can also be a hindrance to broadband deployment because it is more expensive to deploy broadband technologies in a mountainous or heavily forested area. An additional added cost factor for remote areas can be the expense of "backhaul"

(e.g., the "middle mile") which refers to the installation of a dedicated line which transmits a signal to and from an Internet backbone which is typically located in or near an urban area.

[14] Gillett, Sharon E., Massachusetts Institute of Technology, *Measuring Broadband's Economic Impact*, report prepared for the Economic Development Administration, U.S. Department of Commerce, February 28, 2006 p. 4.

[15] Crandall, Robert, William Lehr, and Robert Litan, *The Effects of Broadband Deployment on Output and Employment: A Cross-sectional Analysis of U.S. Data*, June 2007, 20 pp. Available at http://www3.brookings.edu/ views/papers/crandall/200706litan.pdf.

[16] International Telecommunications Union, *Economies by broadband penetration, 2007.* Available athttp://www.itu.int/ITU-D/ict/statistics/at_glance/top20_broad_2007.html.

[17] OECD, *OECD Broadband Statistics, June 2008*. Available at http://www.oecd.org/sti/ict/broadband.

[18] OECD, Directorate for Science, Technology and Industry, *The Development of Broadband Access in OECD Countries*, October 29, 2001, 63 pp. For a comparison of government broadband policies, also see OECD, Directorate for Science, Technology and Industry, *Broadband Infrastructure Deployment: The Role of Government Assistance*, May 22, 2002, 42 pp.

[19] See Turner, Derek S., Free Press, *Broadband Reality Check II: The Truth Behind America's Digital Divide*, August2006, pp 8-11. Available at http://www.freepress.net/files/bbrc2-final.pdf; and Turner, Derek S., Free Press, *'Shooting the Messenger' Myth vs. Reality: U.S. Broadband Policy and International Broadband Rankings*, July 2007, 25 pp., available at http://www.freepress.net/files/shooting_the_messenger.pdf.

[20] National Telecommunications and Information Administration, *Fact Sheet: United States Maintains Information and Communication Technology (ICT) Leadership and Economic Strength*, at http://www.ntia.doc.gov/ntiahome/press/2007/ICTleader_042407.html.

[21] See Wallsten, Scott, Progress and Freedom Foundation, *Towards Effective U.S. Broadband Policies*, May 2007, 19pp. Available at http://www.pff.org/issues-pubs/pops/pop14.7usbroadbandpolicy.pdf.
Also see Ford, George, Phoenix Center, *The Broadband Performance Index: What Really Drives Broadband Adoption Across the OECD?*, Phoenix Center Policy Paper Number 33, May 2008, 27 pp; available at http://www.phoenix-center.org/pcpp/PCPP33Final.pdf.

[22] See price and services and speed data on OECD Broadband Portal, available at http://www.oecd.org/sti/ict/broadband; Turner, Derek S., Free Press, *Broadband Reality Check II: The Truth Behind America's Digital Divide*, August 2006, pp 5-9; Kende, Michael, Analysis Consulting Limited, *Survey of International Broadband Offerings*, October 4, 2006, 12 p, available at http://www.analysys.com/pdfs/BroadbandPerformanceSurvey.pdf; and Atkinson, Robert D., The International Technology and Innovation Foundation, *Explaining International Broadband Leadership*, May 2008, 108 p, available at http://www.itif.org/files/Explaining BBLeadership.pdf.

[23] Federal Communications Commission, *Notice of Inquiry*, "Concerning the Deployment of Advanced Telecommunications Capability to All Americans in a Reasonable and Timely Fashion, and possible Steps to Accelerate Such Deployment Pursuant to Section 706 of the Telecommunications Act of 1996," FCC 04-55, March 17, 2004, p. 6.

[24] FCC News Release, *FCC Improves Data Collection to Monitor Nationwide Broadband Rollout*, November 9, 2004.

[25] Available at http://hraunfoss.fcc.gov/edocs_public/attachmatch/DOC-254115A1.pdf. U.S. Government Accountability Office, *Broadband Deployment is Extensive throughout the United States, but It Is Difficult to Assess the Extent of Deployment Gaps in Rural Areas*, GAO-06-426, May 2006, p. 3.

[26] Federal Communications Commission, *Notice Proposed Rulemaking*, "Development of Nationwide Broadband Data to Evaluate Reasonable and Timely Deployment of Advanced Services to All Americans, Improvement of Wireless Broadband Subscribership Data, and Development of Data on Interconnected Voice Over Internet Protocol (VoIP)Subscribership," WC Docket No. 07-38, FCC 07-17, released April 16, 2007, 56 pp.

[27] FCC, News Release, "FCC Expands, Improves Broadband Data Collection," March 19, 2008. Available athttp://hraunfoss.fcc.gov/edocs_public/attachmatch/DOC-280909A1.pdf.

[28] Testimony of Brian Mefford, President and CEO, Connected Nation, Inc., before the Senate Committee on Commerce, Science and Transportation, April 24, 2007. Available at http://commerce.senate.gov/public/_files/ DC_Committeetestimony_04_23_07.pdf.

[29] Federal Communications Commission, *Fifth Report*, "In the Matter of Inquiry Concerning the Deployment of Advanced Telecommunications

Capability to All Americans in a Reasonable and Timely Fashion, and Possible Steps to Accelerate Such Deployment Pursuant to Section 706 of the Telecommunications Act of 1996," GN Docket No. 07-45, FCC 08-88, Adopted March 19, 2008, Released June 12, 2008. 76 pp. Available at http://hraunfoss.fcc.gov/ edocs_public/attachmatch/FCC-08-88A1.pdf.

[30] Ibid., pp. 5, 7.

[31] See White House, A New Generation of American Innovation, April 2004. Available at http://www.whitehouse.gov/infocus/technology/ economic_policy200404/innovation.pdf.

[32] U.S. Department of Commerce, National Telecommunications and Information Administration, Networked Nation: Broadband in America 2007, January 2008, p. I. Available at http://www.ntia.doc.gov/ report s/2008/NetworkedNationBroadbandinAmerica2007.pdf.

[33] NTIA, *Press Release*, "Gutierrez Hails Dramatic U.S. Broadband Growth," January 31, 2008. Available at http://www.ntia.doc.gov/ ntiahome/press/2008/NetworkedNation_013108.html.

[34] Office of the President-Elect, *Technology Agenda*, available at http://change.gov/agenda/technology_agenda.

[35] Barack Obama, *Connecting and Empowering All Americans Through Technology and Innovation*, 2008, available at http://obama. 3cdn.net/780e0e91ccb6cdbf6e_6udymvin7.pdf.

[36] See CRS Report R40436, *Broadband Infrastructure Programs in the American Recovery and Reinvestment Act*, by Lennard G. Kruger.

[37] The section on universal service was prepared by Angele Gilroy, Specialist in Telecommunications, Resources, Science and Industry Division. For more information on universal service, see CRS Report RL33979, *Universal Service Fund: Background and Options for Reform*, by Angele A. Gilroy.

[38] Communications Act of 1934, As Amended, Title I sec.1 [47 U.S.C. 151].

[39] Many states participate in or have programs that mirror FCC universal service mechanisms to help promote universal service goals within their states.

[40] Additional FCC policies such as rate averaging and pooling have also been implemented to assist high cost carriers.

[41] For additional information on this program, including funding commitments, see the E-rate website: http://www.universal service .org/sl/.

[42] For additional information on this program, including funding commitments, see the RHCD website :http://www.universal service .org/rhc/.

[43] For additional information on the TDF fund and TDF Foundation see the TDF website at http://www.tdfund.com.

[44] The Joint Board recommended that the definition of those services that qualify for universal service support be expanded and that the nation's communications goals include the universal availability of: mobility services (i.e., wireless voice); broadband Internet services; and voice services at affordable and comparable rates for all rural and non-rural areas. For a copy of this recommendation see http://hraunfoss.fcc .gov/edocs_public/attachmatch/FCC-07J-4A1.pdf.

[45] See CRS Report R40149, *Infrastructure Programs: What's Different About Broadband?*, by Charles B. Goldfarb and Lennard G. Kruger, p. 21.

[46] For information on existing broadband programs at RUS, see: CRS Report RL33816, *Broadband Loan and Grant Programs in the USDA's Rural Utilities Service*, by Lennard G. Kruger.

In: Computer Networks
Editor: Raymond S. Grigello

ISBN: 978-1-61209-596-7

Chapter 3

BROADBAND INFRASTRUCTURE PROGRAMS IN THE AMERICAN RECOVERY AND REINVESTMENT ACT*

Lennard G. Kruger

ABSTRACT

The American Recovery and Reinvestment Act (ARRA, P.L. 111-5) provides \$7.2 billion primarily for broadband grant programs to be administered by two separate agencies: the National Telecommunications and Information Administration (NTIA) of the Department of Commerce (DOC) and the Rural Utilities Service (RUS) of the U.S. Department of Agriculture (USDA). Of the \$7.2 billion total, the ARRA provides \$4.7 billion to establish a Broadband Technology Opportunities Program (BTOP) at NTIA, and \$2.5 billion as additional funding for broadband grant, loan, and loan guarantee programs at RUS. Broadband grants and loans funded by the ARRA are competitive and applicants must apply directly to NTIA and RUS. The NTIA appropriation also includes \$350 million for a national broadband inventory map, funding for the Broadband Data Improvement Act (P.L. 110-385), and funding to be transferred to the Federal Communications Commission (FCC) to develop a national broadband plan.

The unprecedented scale and scope of the ARRA broadband programs, coupled with the short time frame for awarding grants, presents

* This is an edited, excerpted and augmented edition of a CRS Report, Report R40436 dated March 20, 2009.

daunting challenges with respect to program implementation as well as Congressional oversight. Congress is closely monitoring how equitably and effectively broadband grants are allocated among states and the various stakeholders, and to what extent the programs fulfill the goals of short term job creation and the longer term economic benefits anticipated from improved broadband availability, access, and adoption. A continuing issue is how to strike a balance between providing federal assistance for unserved and underserved areas where the private sector may not be providing acceptable levels of broadband service, while at the same time minimizing any deleterious effects that government intervention in the marketplace may have on competition and private sector investment.

Implementation decisions made by NTIA and RUS could have a significant impact on how the program is shaped and targeted, and the extent to which the program meets the goals and purposes set forth by the ARRA. Some implementation issues expected to be addressed include defining "underserved" and "unserved" areas with respect to broadband service, defining "nondiscrimination and network interconnection obligations," defining "broadband," the role of the states, coordination between federal agencies, broadband data collection, and evaluation and transparency.

INTRODUCTION

Broadband infrastructure refers to networks of deployed telecommunications equipment and technologies necessary to provide high-speed Internet access and other advanced telecommunications services for private homes, businesses, commercial establishments, schools, and public institutions. In the United States, broadband infrastructure is constructed, operated, and maintained primarily by the private sector, including telephone, cable, satellite, wireless, and other information technology companies. Currently deployed broadband technologies include cable modem, DSL (copper wire), wireless systems (mobile and fixed), fiber, and satellite. Although broadband is deployed by private sector providers, federal and state regulation of the telecommunications industry as well as government financial assistance programs can have a significant impact on private sector decisions to invest in and deploy broadband infrastructure, particularly in underserved and unserved areas of the nation.

The American Recovery and Reinvestment Act (ARRA, P.L. 111-5) provides $7.2 billion primarily for broadband grant programs to be administered by two separate agencies: the National Telecommunications and

Information Administration (NTIA) of the Department of Commerce (DOC) and the Rural Utilities Service (RUS) of the U.S. Department of Agriculture (USDA). Of the $7.2 billion total, the ARRA provides $4.7 billion to establish a Broadband Technology Opportunities Program (BTOP) at NTIA, and $2.5 billion for broadband grant, loan, and loan guarantee programs at RUS. The ARRA also directs the Federal Communications Commission (FCC) to develop a national broadband strategy. In comparison with previously existing federal broadband programs in the United States, [1] the broadband grant programs established and funded by P.L. 111-5 are unprecedented in scale and scope.

The impetus behind broadband provisions in the ARRA was two-fold: in the short term, to create jobs through the construction and deployment of broadband infrastructure, and in the long term, to address concerns over economic and societal impacts of inadequate broadband availability, access, and adoption, particularly in rural and lower-income areas of the nation [2]. The unprecedented scale and scope of the ARRA broadband programs, coupled with the short time frame for awarding grants, presents daunting challenges with respect to program implementation as well as Congressional oversight.

American Recovery and Reinvestment Act of 2009, P.L. 111-5

In December 2008, leadership in the House and Senate, as well as the Obama transition team, announced their intention to include a broadband component in the infrastructure portion of the economic stimulus package. At the same time, numerous interested parties, including broadband equipment manufacturers; large, mid-sized, and small wireline and wireless service providers; satellite operators; telecommunications unions; consumer groups; education groups; public safety organizations; think tanks; and others unveiled a multitude of specific proposals for government support of broadband infrastructure [3].

The House and Senate approved the Conference Report on H.R. 1 (H.Rept. 111-16) on February 13, 2009. On February 17, 2009, President Obama signed P.L. 111-5, the American Recovery and Reinvestment Act (ARRA). Broadband provisions of the ARRA provided a total of $7.2 billion, primarily for broadband grants. The total consists of $4.7 billion to NTIA/DOC for a newly established Broadband Technology Opportunities

Program (BTOP) and $2.5 billion to RUS/USDA broadband grant, loan, and loan guarantee programs [4].

NTIA/DOC

Of the $4.7 billion appropriated to NTIA:

- $4.35 billion is directed to a competitive broadband grant program, of which not less than $200 million shall be available for competitive grants for expanding public computer center capacity (including at community colleges and public libraries); not less than $250 million to encourage sustainable adoption of broadband service; and $10 million transferred to the DOC Office of Inspector General for audits and oversight;
- $350 million is directed for funding the Broadband Data Improvement Act (P.L. 110-385) and for the purpose of developing and maintaining a broadband inventory map, which shall be made accessible to the public no later than two years after enactment; and
- Funds deemed necessary and appropriate by the Secretary of Commerce, in consultation with the FCC, may be transferred to the FCC for the purposes of developing a national broadband plan, which shall be completed one year after enactment.

The Broadband Technology Opportunities Program within NTIA is authorized by Division B, Title VI of the ARRA. Specific implementation requirements and guidelines for the new NTIA broadband grants are as follows:

- Establishes a "national broadband service development and expansion program" with purposes to include providing access to broadband service to consumers residing in unserved and underserved areas; providing broadband education, awareness, training, access, equipment and support to various institutions; improving access to, and use of, broadband service by public safety agencies; and stimulating demand for broadband, economic growth, and job creation.

- Directs NTIA to consult with each state to identify unserved and underserved areas (with respect to access to broadband service) as well as the appropriate allocation of grant funds within that state.
- Directs NTIA, to the extent practical, to award not less than one grant in each state.
- Does not define "unserved area," "underserved area," and "broadband." The Conferees instructed NTIA to coordinate its understanding of these terms with the FCC, and in defining "broadband service" to take into consideration technical differences between wireless and wireline networks and to consider the actual speeds these networks are able to deliver to consumers under a variety of circumstances.
- Directs NTIA, in coordination with the FCC, to publish "non-discrimination and network interconnection obligations" that shall be contractual conditions of awarded grants, and specifies that these obligations should adhere, at a minimum, to the FCC's broadband principles to promote the openness and interconnected nature of the Internet (FCC 05-151, adopted August 5, 2005) [5].
- Directs NTIA, when considering applications for grants, to consider whether the project will provide the greatest broadband speed possible to the greatest population of users in the area. There are no specific speed thresholds that applicants must meet to be eligible for a grant. The Conferees acknowledged that while speed thresholds could have the unintended effect of thwarting broadband deployment in some areas, deploying next-generation speeds would likely result in greater job creation and job preservation. NTIA is instructed to "seek to fund, to the extent practicable, projects that provide the highest possible, next- generation broadband speeds to consumers."
- Defines entities eligible for grants as: a state or political division thereof; the District of Columbia; a territory or possession of the United States; an Indian tribe or native Hawaiian organization; a nonprofit foundation, corporation, institution or association; or any other entity, including a broadband service or infrastructure provider, that NTIA finds by rule to be in the public interest.
- Requires NTIA to consider whether a grant applicant is a socially and economically disadvantaged small business as defined under the Small Business Act.

- Directs NTIA to ensure that all awards are made before the end of FY20 10. Grantees will be required to substantially complete projects within two years after the grant is awarded.
- Directs that the federal share of any project cannot exceed 80% unless the applicant petitions NTIA and demonstrates financial need.
- Directs that grant applicants must demonstrate that the grant project would not have been implemented during the grant period without federal grant assistance.

RUS/USDA

The $2.5 billion appropriated to RUS is designated as additional amounts for RUS grant, loan, and loan guarantee programs. The ARRA does not specify how the $2.5 billion is to be divided between grants and loans. Regarding projects applying for ARRA funding, the law states that:

- at least 75% of the area to be served by a project receiving these funds shall be in a rural area without sufficient access to high-speed broadband service to facilitate economic development, as determined by the Secretary of Agriculture;
- priority shall be given to projects that will deliver end users a choice of more than one broadband service provider;
- priority shall be given to projects that provide service to the highest proportion of rural residents that do not have access to broadband service;
- priority shall be given to borrowers and former borrowers of rural telephone loans;
- priority shall be given to projects demonstrating that all project elements will be fully funded, that can commence promptly, and that can be completed; and
- no area of a project may receive funding to provide broadband service under the Broadband Technology Opportunities Program at NTIA/DOC.

IMPLEMENTATION OF ARRA BROADBAND PROGRAMS

Broadband grants funded by the ARRA are competitive and applicants must apply directly to NTIA and RUS. Rules, regulations, and application guidelines for those interested in applying for broadband grants are in development by the agencies. Websites tracking the latest ARRA broadband program developments are located at NTIA, [6] RUS, [7] and the FCC [8].

Although grants are competitive and available directly from the federal agencies, the states will also play a role, particularly with respect to the BTOP at NTIA. The ARRA directs NTIA to consult with each state to identify unserved and underserved areas (with respect to access to broadband service) as well as the appropriate allocation of grant funds within that state. However, final award decisions reside with the federal agencies.

On March 10, 2009, a public meeting was held by NTIA, RUS, and the FCC. NTIA officials indicated that they anticipate holding three separate broadband grant rounds, the first between April and June 2009, the second between October and December 2009, and the third between April and June 2010. Preceding each round a Notice of Funding Availability (NOFA) will be released providing application details.

RUS officials indicated that they would use the budget authority provided by the ARRA to support a program of grants, loans, loan-guarantees, and possibly loan-grant combinations. RUS anticipates at least three application rounds and expects to issue its first NOFA between April and June.

Meanwhile, the FCC announced it plans to issue a Notice of Inquiry on April 8, 2009, to gather data, expertise, and public input in preparation for its effort to develop a national broadband strategy as mandated by the ARRA.

ISSUES RELATED TO IMPLEMENTATION

The Broadband Technology Opportunities Program (BTOP) is newly authorized and established by the ARRA. The law gives NTIA considerable flexibility to implement the BTOP. According to the Conference Report:

> The Conferees intend that the NTIA has discretion in selecting the grant recipients that will best achieve the broad objectives of the program. The Conferees also intend that the NTIA select grant recipients that it judges will best meet the broadband access needs of the area to be served, whether by a

wireless provider, a wireline provider, or any provider offering to construct last-mile, middle-mile, or long haul facilities.

Implementation decisions made by NTIA could have a significant impact on how the program is shaped and targeted, and the extent to which the program meets the goals and purposes set forth by the ARRA. NTIA and RUS have prepared a joint request for information (RFI) and notice of public meetings designed to gather public input into many of the implementation decisions which the agencies will make as they develop rules and regulations for the grant program [9]. A series of public meetings are being held in March 2009. The RFI is soliciting comments from all interested parties on the following topics:

- the purposes of the BTOP program;
- the role of the states;
- eligible grant recipients;
- the establishment of selection criteria for grant awards;
- grant mechanics;
- grants for expanding public computer center capacity;
- grants for innovative programs to encourage sustainable adoption of broadband service;
- broadband mapping;
- financial contributions by grant applicants;
- timely completion of proposals;
- coordination between BTOP and the RUS grant program;
- how terms set out in relevant sections of the ARRA should be defined;
- how the success of the BTOP program should be measured;
- any other issues NTIA should consider in creating the BTOP;
- the most effective ways RUS could offer broadband funds;
- how RUS and NTIA can best align their activities;
- how RUS can evaluate whether a particular level of broadband access and service is needed to facilitate economic development;
- how RUS should consider priorities set out in the ARRA in selecting applications; and
- what benchmarks should be used to determine the success of RUS ARRA broadband activities.

Below is a discussion of selected issues that will be addressed as implementation details of the ARRA broadband programs are developed.

Defining "Underserved" and "Unserved"

As specified in the ARRA, the purpose of BTOP is to provide broadband service to consumers residing in unserved and underserved areas of the United States. The issue of which areas should be defined as "underserved" with respect to broadband service has long been controversial. There is no generally accepted definition of "underserved." Factors such as a minimal number of existing providers, a lack of adequate market competition, unaffordable consumer prices for existing broadband service, or substandard download and upload available speeds may singularly or in combination lead some to define an area as "underserved." The definition of "unserved" is also not uniformly accepted. For example, should unserved be defined only as an area with no terrestrial (nonsatellite) broadband service, or should areas with some terrestrial but no mobile wireless service also be considered "unserved?"

The ARRA does not define either "unserved" or "underserved." The law directs NTIA to consider whether a grant application would increase broadband affordability and subscribership, and provide the greatest broadband speeds possible to the greatest population of users in the area served. The ARRA directs NTIA to consult with the states (plus the District of Columbia and the territories) to identify unserved and underserved areas within that state. The Conferees instructed NTIA to coordinate its understanding of the terms "unserved area" and "underserved area" with the FCC.

In approaching an understanding of these terms, the NTIA (and the states with which the NTIA will consult on this issue) must balance competing policy concerns, particularly when developing or embracing a definition of "underserved." For example, too narrow a definition may make it more difficult for rural areas in need of adequate broadband service to receive grants. On the other hand, too broad a definition of "underserved" may inappropriately confront existing broadband providers with government-funded competitors and may divert funding from projects in unserved areas with no broadband service whatsoever. Both NTIA and the states will likely seek to strike a balance between those two competing concerns.

Defining "Non-Discrimination and Network Interconnection Obligations"

Congressional policymakers continue to debate and consider whether laws or regulations are needed to ensure the "openness" of the Internet with respect to both content and access [10]. The debate over nondiscrimination (also commonly referred to as "net neutrality," "open access," and "network management") has shifted to a debate over the extent to which nondiscrimination requirements or standards should be imposed on broadband networks funded by BTOP. The ARRA directs NTIA, in coordination with the FCC, to publish "non-discrimination and network interconnection obligations that shall be contractual conditions of grants awarded." The ARRA says that these obligations, at a minimum, should adhere to the principles contained in the FCC's broadband policy statement (FCC 05-15, adopted August 5, 2005) as follows:

> To encourage broadband deployment and preserve and promote the open and interconnected nature of the public Internet, consumers are entitled to access the lawful Internet content of their choice.
>
> To encourage broadband deployment and preserve and promote the open and interconnected nature of the public Internet, consumers are entitled to run applications and use services of their choice, subject to the needs of law enforcement.
>
> To encourage broadband deployment and preserve and promote the open and interconnected nature of the public Internet, consumers are entitled to connect their choice of legal devices that do not harm the network.
>
> To encourage broadband deployment and preserve and promote the open and interconnected nature of the public Internet, consumers are entitled to competition among network providers, application and service providers, and content providers [11].

In developing nondiscrimination and interconnection obligations for funded projects, NTIA and the FCC face the challenge of ensuring the "openness" of federally funded broadband networks, while at the same time minimizing regulatory burdens on prospective grantees that, some say, [12] may constitute a disincentive for some entities to apply. How NTIA and the FCC choose to balance these two considerations could prove highly controversial as the BTOP implementation rules and regulations are developed.

Defining Broadband

The term "broadband" is typically characterized or defined by minimum download and upload speeds, specific technologies (i.e., cable modem, fiber-to-the-home, wifi), or specific applications (e.g., telemedicine, distance learning). The ARRA broadband provisions do not specify minimum download/upload speed thresholds, are technology neutral, and cite a wide variety of applications eligible for funding.

The issue of speed thresholds is particularly controversial. While a high speed threshold has the benefit of encouraging the construction of next generation networks (such as fiber or next generation cable systems), it also runs the risk of excluding current generation technologies that may be uniquely suitable for some unserved or underserved areas. The Conferees acknowledged this dilemma, stating in the Conference Report that while speed thresholds could have the unintended effect of thwarting broadband deployment in some areas, deploying next-generation speeds would likely result in greater job creation and job preservation. The Conferees instructed NTIA to "seek to fund, to the extent practicable, projects that provide the highest possible, next- generation broadband speeds to consumers"[13]. Thus, NTIA has the flexibility to balance the sometimes competing goals of constructing next generation networks with providing broadband to unserved and underserved areas.

Role of the States

While the BTOP grants are competitive and will be awarded directly by NTIA, the states are expected to play a significant role. The ARRA directs NTIA to consult with each state to identify unserved and underserved areas (with respect to access to broadband service) as well as the appropriate allocation of grant funds within that state. States themselves (as well as municipalities) are eligible to apply for broadband grants, and the ARRA specifies that NTIA, to the extent practical, shall award not less than one grant to an entity within each state.

Regarding NTIA consultation with the states, the Conferees expressed the following:

> The Conferees recognize that States have resources and a familiarity with local economic, demographic, and market conditions that could

> contribute to the success of the broadband grant program. States are encouraged to coalesce stakeholders and partners, assess community needs, aggregate demand for services, and evaluate demand for technical assistance. The Conferees therefore expect and intend that the NTIA, at its discretion, will seek advice and assistance from the States in reviewing grant applications, as long as the NTIA retains the sole authority to approve the awards. The Conferees further intend that the NTIA will, in its discretion, assist the States in post-grant monitoring to ensure that recipients comply fully with the terms and conditions of their grants [14].

An issue will likely be to what extent the NTIA follows the recommendations of the states with respect to award decisions. 15 States vary widely with respect to their own broadband programs and initiatives. Some states have embarked on state-wide broadband strategies and have been extremely active in mapping broadband availability and identifying unserved and underserved areas, while other states have not yet begun such an effort [16].

Eligibility

The ARRA defines eligible entities for BTOP grants as a state or political division thereof; the District of Columbia; a territory or possession of the United States; an Indian tribe or native Hawaiian organization; a nonprofit foundation, corporation, institution or association; or any other entity, including a broadband service or infrastructure provider, that NTIA finds by rule (required to be technologically neutral) to be in the public interest.

NTIA is thus directed to set the parameters of eligibility beyond states, political subdivisions, and nonprofit organizations. The issue is the extent to which eligibility will be extended to private sector for-profit broadband providers, be they large or small, incumbents or new entrants. According to the ARRA Conference Report, it was the intent of the Conferees that as many entities as possible be eligible to apply for a grant, including wireless carriers, wireline carriers, backhaul providers, satellite carriers, public-private partnerships, and tower companies.

Coordination between Agencies

Three federal agencies—NTIA, RUS, and the FCC—will be implementing the broadband provisions of the ARRA. Coordination between those agencies will likely be an important factor in ensuring that broadband programs meet the goals of the ARRA. Specific coordination challenges include ensuring that NTIA's BTOP grants and the RUS grant and loan programs are complementary and not duplicative; ensuring that the FCC has appropriate input into the design and implementation of broadband grant and data programs; and coordinating the FCC's national broadband plan (as required by the ARRA) with RUS and NTIA.

Broadband Data Gathering

There is widespread agreement that data regarding broadband deployment in the United States are inadequate and that policymakers have an incomplete picture of where broadband service is available (and at what speeds and prices). Broadband data are important, because the more detailed and granular broadband data are, the more effectively government can direct broadband assistance to areas with the greatest need.

The ARRA addressed broadband data by designating $350 million for funding the Broadband Data Improvement Act (P.L. 110-385) and for the purpose of developing and maintaining a national broadband inventory map. The Broadband Data Improvement Act (P.L. 110-385) was signed into law on October 10, 2008, and requires the FCC to collect demographic information on unserved areas, data comparing broadband service with 75 communities in at least 25 nations abroad, and data on consumer use of broadband. The act also directs the Census Bureau to collect broadband data, the Government Accountability Office to study broadband data metrics and standards, and the Department of Commerce to provide grants supporting state broadband data, mapping, and planning initiatives.

Regarding the inventory map, the ARRA directs NTIA to develop and maintain a comprehensive nationwide inventory map of existing broadband service capability and availability in the United States that depicts the geographic extent to which broadband service capability is deployed and available from a commercial provider or public provider throughout each state. Not later than two years after enactment of the ARRA, the NTIA is directed to make the national inventory map available online to the public in a form that is interactive and searchable.

A continuing and controversial issue related to broadband data is striking a balance between making available broadband deployment data to the public that is sufficiently detailed to be useful, without revealing what some providers may consider to be proprietary information.

Evaluation and Transparency

Given the large amounts of grant money to be awarded within tight deadlines (by September 30, 2010, for the BTOP grants), there is considerable interest in the issue of transparency and how the programs will be evaluated and monitored in order to avoid waste, fraud, and abuse. To address this issue, the ARRA:

- requires the Secretary of Agriculture to submit a report to the House and Senate Appropriations Committees on planned spending and actual obligations describing the use of ARRA funds ($2.5 billion) for the RUS broadband programs not later than 90 days after enactment, and quarterly thereafter until all funds are obligated;
- transfers $10 million to the Department of Commerce Office of Inspector General for audits and oversight of funds provided for the Broadband Technology Opportunities Program;
- directs NTIA to report every 90 days on the status of BTOP to the House and Senate Appropriations Committees, the House Committee on Energy and Commerce, and the Senate Committee on Commerce, Science and Transportation;
- directs NTIA to require grant recipients to file quarterly reports (which will be publicly available) on the grantee's use of the grant money and progress on fulfilling the objectives for which the funds were granted;
- authorizes NTIA, if it chooses, to establish additional reporting and information requirements for any grant recipient;
- authorizes NTIA, in addition to other authority under applicable law, to deobligate awards to grantees that demonstrate an insufficient level of performance, or wasteful or fraudulent spending, as defined in advance by NTIA, and award these funds competitively to new or existing applicants; and

- directs NTIA to create and maintain a fully searchable database, accessible on the Internet at no cost to the public, that contains at least a list of each entity that has applied for a grant, a description of each application, the status of each application, the name of each entity receiving funds, the purpose for which the entity is receiving funds, each quarterly report submitted by the entity, and other information sufficient to allow the public to understand and monitor grants awarded under the program.

An issue is whether the transparency and evaluation requirements as set forth in the ARRA and other law are sufficient, or whether additional reporting and information requirements will be adopted.

CONCLUDING OBSERVATIONS

The broadband programs in the ARRA, funded at $7.2 billion, are unprecedented in scope and scale compared with previously existing federal broadband assistance programs. Policy decisions made by NTIA, RUS, and the FCC could have major impacts on the implementation of the program and the extent to which it meets the goals set by Congress for short-term job creation and long-term improvement of the nation's broadband infrastructure.

The ARRA broadband provisions are only one component in the nation's overall broadband strategy. Among other issues which may be addressed as part of a national broadband policy (likely to be formulated by the Administration and the FCC) are universal service reform, tax incentives to encourage private sector broadband rollout, and spectrum policy to spur roll-out of wireless broadband services. As Congress continues to monitor broadband stimulus programs, while considering various additional options for encouraging broadband deployment and adoption, a key issue is how to strike a balance between providing federal assistance for unserved and underserved areas where the private sector may not be providing acceptable levels of broadband service, while at the same time minimizing any deleterious effects that government intervention in the marketplace may have on competition and private sector investment.

References

[1] See CRS Report RL30719, Broadband Internet Access and the Digital Divide: Federal Assistance Programs, by Lennard G. Kruger and Angele A. Gilroy.

[2] See Ibid., pp. 1-4.

[3] See CRS Report R40 149, Infrastructure Programs: What's Different About Broadband?, by Charles B. Goldfarb and Lennard G. Kruger.

[4] For information on stimulus funding directed to the existing broadband programs at RUS, see CRS Report RL33816, Broadband Loan and Grant Programs in the USDA's Rural Utilities Service, by Lennard G. Kruger.

[5] See CRS Report RS22444, Net Neutrality: Background and Issues, by Angele A. Gilroy.

[6] http://www.ntia.doc.gov/broadbandgrants.

[7] http://www.usda.gov/rus/telecom/index.htm.

[8] http://www.fcc.gov/recovery/broadband/.

[9] Department of Commerce, National Telecommunications and Information Administration, and Department of Agriculture, Rural Utilities Service "American Recovery and Reinvestment Act of 2009 Broadband Initiatives," 74 Federal Register 10716-10721, March 12, 2009.

[10] See CRS Report RS22444, Net Neutrality: Background and Issues, by Angele A. Gilroy.

[11] FCC, Policy Statement on Broadband Internet Access, FCC 05-151, adopted August 5, 2005, available at http://hraunfoss.fcc.gov/edocs_public/attachmatch/FCC-05-151A1.pdf. The FCC principles are not rules – rather they are intended as general principles to be incorporated in FCC's ongoing policymaking activities.

[12] Stephanie Condon, "Telecoms Oppose Tighter Net Neutrality Rules for Stimulus Funds," CNET News, February 26, 2009.

[13] U.S. Congress, Conference Report to Accompany H.R. 1, 111th Cong., 1st sess., February 12, 2009, H.Rept. 111-16 (Washington: GPO, 2009), p. 775.

[14] Ibid.

[15] Some states have already set up their own ARRA websites and have begun soliciting proposals for grant funding, including broadband projects. See http://www.recovery.gov/?q=content/state-recovery-page.

[16] For more information, see State Broadband Initiatives: A Summary of State Programs Designed to Stimulate Broadband Deployment and

Adoption, A Joint Report of the Alliance for Public Technology and the Communications Workers of America, July 2008, 54 pages. State program database available at http://www.speedmatters.org/ statepolicy.

In: Computer Networks
Editor: Raymond S. Grigello

ISBN: 978-1-61209-596-7

Chapter 4

COMPREHENSIVE NATIONAL CYBERSECURITY INITIATIVE: LEGAL AUTHORITIES AND POLICY CONSIDERATIONS*

John Rollins and Anna C. Henning

ABSTRACT

Federal agencies report increasing cyber-intrusions into government computer networks, perpetrated by a range of known and unknown actors. In response, the President, legislators, experts, and others have characterized cybersecurity as a pressing national security issue.

Like other national security challenges in the post-9/11 era, the cyber threat is multi-faceted and lacks clearly delineated boundaries. Some cyber attackers operate through foreign nations' military or intelligence-gathering operations, whereas others have connections to terrorist groups or operate as individuals. Some cyber threats might be viewed as international or domestic criminal enterprises.

In January 2008, the Bush Administration established the Comprehensive National Cybersecurity Initiative (the CNCI) by a classified joint presidential directive. The CNCI establishes a multi-pronged approach the federal government is to take in identifying current and emerging cyber threats, shoring up current and future

* This is an edited, excerpted and augmented edition of a CRS Report, Report R40447 dated March 10, 2009.

telecommunications and cyber vulnerabilities, and responding to or proactively addressing entities that wish to steal or manipulate protected data on secure federal systems. On February 9, 2009, President Obama initiated a 60-day interagency cybersecurity review to develop a strategic framework to ensure the CNCI is being appropriately integrated, resourced, and coordinated with Congress and the private sector.

In response to the CNCI and other proposals, questions have emerged regarding: (1) the adequacy of existing legal authorities—statutory or constitutional—for responding to cyber threats; and (2) the appropriate roles for the executive and legislative branches in addressing cybersecurity. The new and emerging nature of cyber threats complicates these questions. Although existing statutory provisions might authorize some modest actions, inherent constitutional powers currently provide the most plausible legal basis for many potential executive responses to national security related cyber incidences. Given that cyber threats originate from various sources, it is difficult to determine whether actions to prevent cyber attacks fit within the traditional scope of executive power to conduct war and foreign affairs. Nonetheless, under the Supreme Court jurisprudence, it appears that the President is not prevented from taking action in the cybersecurity arena, at least until Congress takes further action. Regardless, Congress has a continuing oversight and appropriations role. In addition, potential government responses could be limited by individuals' constitutional rights or international laws of war. This report discusses the legal issues and addresses policy considerations related to the CNCI.

INTRODUCTION

Cybersecurity has been called "one of the most urgent national security problems facing the new administration." [1] Cyber and telecommunications activities are sometimes conflated to indicate the same meaning or capability. One might distinguish the term cyber from that of telecommunications with the former being the data or applications residing on the latter which is the electronic medium in which the activity occurs. Electronic information systems, also termed "information infrastructures," now support a wide range of security and economic assets in the public and private sectors.

Such systems have been successfully infiltrated in recent years by a range of attackers, some of whom are suspected to have been working in coordination with foreign military organizations or (foreign) state intelligence services. Thus, like the changing nature of U.S. enemies in the post- 9/11 environment, the nature of military and economic vulnerabilities has changed:

intelligence- gathering battles in cyberspace now also play a crucial role in national security.

In January 2008, the Bush Administration initiated the Comprehensive National Cybersecurity Initiative (the CNCI) to make the United States more secure against cyber threats. The Homeland Security Presidential Directive 23 and National Security Presidential Directive 54 establishing the CNCI are classified. Some details of the Initiative have been made public in Departmental press releases, speeches by executive branch leaders, and analysis and insight offered by individuals that follow cyber security and terrorism related issues. The CNCI "establishes the policy, strategy, and guidelines to secure federal systems." [2] The CNCI also delineates "an approach that anticipates future cyber threats and technologies, and requires the federal government to integrate many of its technical and organizational capabilities to better address sophisticated threats and vulnerabilities." [3] Subsequent to the issuance of the classified directives, congressional committees have held hearings regarding the CNCI and heard testimony from a commission established to address necessary cybersecurity reforms [4].

In a speech during his presidential campaign, President Obama promised to "make cyber security the top priority that it should be in the 21st century ... and appoint a National Cyber Advisor who will report directly" to the President [5] Although the Obama Administration might craft a new approach to cybersecurity, some experts have urged the new administration to build on the CNCI, which they note is a "major step toward improving federal cybersecurity" [6]. On February 9, 2009, President Obama directed a 60-day interagency cybersecurity review to develop a strategic framework to ensure the CNCI is being appropriately integrated, resourced, and coordinated with Congress and the private sector [7].

The new administration's focus on cybersecurity would continue recent emphasis on the issue by the executive and legislative branches. This recent focus emerged partly in response to events such as attacks by outside hackers against a Pentagon computer network and the CyberWar against Estonia, which garnered significant media attention. Agency reports of large numbers of attempts to infiltrate government cyberspace have also prompted action. Both the high-profile attacks and more routine infiltrations have shed light on the vulnerability of critical information infrastructures. For example, the Defense Science Board noted that the U.S. military's information infrastructure is the "Achilles' heel of our otherwise overwhelming military might"[8].

BACKGROUND ON CYBER THREATS AND CALLS FOR EXECUTIVE ACTION

Threats to the U.S. cyber and telecommunications infrastructure are constantly increasing [9] and evolving as are the entities that show interest in using a cyber-based capability to harm the nation's security interests [10] Concerns have been raised since the 1990s regarding the use of the internet and telecommunications components to cause harm to the nation's security interests. Activities producing undesirable results include unauthorized intrusion to gain access and view protected data, stealing or manipulating information contained in various databases, and attacks on telecommunications devices to corrupt data or cause infrastructure components to operate in an irregular manner. Of paramount concern to the national and homeland security communities is the threat of a cyber related attack against the nation's critical government infrastructures – "systems and assets, physical or virtual, so vital to the United States that the incapacity or destruction of such systems and assets would have a debilitating impact on security, national economic security, national public health and safety, or any combination of those matters" [11]. Early concerns noted attacks on components of the energy grid, infrastructure control systems, and military equipment as examples of telecommunications based threats to physical infrastructures [12].

In response, the Department of Energy conducted an experiment in 2007 in which the control system of an unconnected generator, containing similar components as that of larger generators connected to many power grids in the nation supplying electricity, was damaged and became inoperable [13]. While data from federal agencies demonstrate that the majority of attempted and successful cyber attacks to date have targeted virtual information resources rather than physical infrastructures, [14] many security experts are concerned that the natural progression of those wishing to harm U.S. security interests will transition from stealing or manipulating data to undertaking action that temporarily or permanently disables or destroys the telecommunication network or affects infrastructure components. Many security observers agree that the United States currently faces a multi-faceted, technologically based vulnerability in that "our information systems are being exploited on an unprecedented scale by state and non-state actors [resulting in] a dangerous combination of known and unknown vulnerabilities, strong adversary capabilities, and weak situational awareness"[15]. This, coupled with security observers' contention that the United States lacks the capability to definitively ascertain perpetrators who might unlawfully access a database or cause harm to a network, leaves the nation increasingly at risk. It also causes acts or discussions related to deterring cyberattacks to be ignored or negated by entities exploiting known or newly found vulnerabilities.

Prominent national security experts have emphasized the vulnerability of U.S. infrastructures. As recently as January 2009, former Director of National Intelligence (DNI) Mike McConnell equated "cyber weapons" with weapons of mass destruction when he expressed concern about terrorists' use of technology to degrade the nation's infrastructure. In distinguishing between individuals gaining access to U.S. national security systems or corporate data for purposes of exploitation for purposes of competitive advantage, former Director McConnell noted that terrorists aim to damage infrastructure and that the "time is not too far off when the level of sophistication reaches a point that there could be strategic damage to the United States."[16]

Similarly, in elaborating on the potential consequences of a cyber attack, newly confirmed DNI Dennis Blair offered the following statement during the Annual Threat Assessment of the Intelligence Community for the Senate Select Committee on Intelligence:

> Growing connectivity between information systems, the Internet, and other infrastructures creates opportunities for attackers to disrupt telecommunications, electrical power, energy pipelines, refineries, financial networks, and other critical infrastructures. Over the past several years we have seen cyber attacks against critical infrastructure abroad, and many of our own infrastructures are as vulnerable as their foreign counterparts. A successful attack against a major financial service provider could severely impact the national economy, while cyber attacks against physical infrastructure computer systems such as this that control power grids or oil refineries have the potential to disrupt services for hours to weeks [17].

Also describing the evolving threat to U.S. security interests from a cyber-facilitated incident, Melissa Hathaway, Senior Advisor to the DNI and Chair of the Nation Cyber Study Group and President Obama's appointee to lead the 60-day interagency strategic cyber review, wrote that "both state and non-state adversaries are targeting our information systems and infrastructure for exploitation and potential disruption or destruction" [18]. During the question and answer period of the most recent DNI Annual Threat Assessment of the Intelligence Community, Director Blair stated that a "cyber capability is not one in which I feel [terrorists] have the skills for the greatest destruction. I think that they have other terrible things they can do to us that they are working on harder, they're better able to do, and they seem to be more motivated to do. So [a cyber terrorist attack is] possible, but I don't think the combination of terror and cyber is the nexus that we are most worried about"[19]. However, threats could originate from foreign military or intelligence operatives rather than from terrorist groups.

In response to reports of the increasing pace and volume of cyber intrusions and a recognition that recent cyber-based threats have compelled the U.S. government to take security related actions that may negatively affect an agency's ability to perform its national security duties, [20] legislators and analysts have expressed concerns that the current statutory framework inadequately addresses modern cybersecurity threats. One prominent voice is the Center for Strategic and International Studies' (CSIS) Commission on Cybersecurity for the 44th President, whose members testified before House and Senate committees and released its formal recommendations in fall 2008. The Commission recommended that federal cyber-crime provisions should be reexamined and that the "President should propose legislation that eliminates the current legal distinction between technical standards for national security systems and civilian agency systems and adopt a risk-based approach to federal computer security" [21]. In addition, it characterized the current statutory framework, particularly the Federal Information Security Management Act, enacted in 2002 to establish agency-level defenses against cyber threats, as too weak to effectively prevent cyber intrusions [22].

Legislators made some attempts during the 110th Congress to strengthen or "modernize" the existing statutory framework. For instance, a bill introduced by Senator Carper, the Federal Information Security Management Act of 2008, [23] would have added a "Chief Information Security Officer" position to supplement the Chief Information Officer position required in each federal agency under the Federal Information Security Management Act of 2002 and the Clinger- Cohen Act of 1996. [24] However, analysts have argued that ultimately, no change to the existing statutory scheme will adequately equip executive agencies to prevent infiltrations into U.S. cyberspace. They argue that "only the White House has the necessary authority and oversight for cybersecurity"[25].

COMPREHENSIVE NATIONAL CYBERSECURITY INITIATIVE AND CONCERNS REGARDING TRANSPARENCY AND EFFECTIVENESS

As of the date of this report, unclassified versions of the January 2008 directives establishing the CNCI have yet to be released. While the Initiative has yet to be legislatively recognized, presidential directives, sometimes considered types of executive orders and visa versa, have the force of law if they are supported by constitutional or statutory authority [26]. Although much remains unknown about the CNCI due to the classified nature of the presidential directives and supporting implementation documents, federal government agency press releases and statements by government officials provide a bit of insight regarding the program. Some security observers are concerned that because the CNCI is focused on developing and adhering to strategies and policies to secure the federal systems, many of which rely on private sector telecommunications networks for service and support, and identifying current and emerging threats and vulnerabilities, it is incumbent on the federal government to improve its coordination activities with non-federal entities and undertake enhanced sharing of timely and relevant cybersecurity related plans and risk data.

Few details have been publicly released regarding the implementation activities or status of CNCI efforts since the establishment of the initiative. According to one media account, Steven Chabinsky, Deputy Director of the Joint Interagency Cyber Task Force for the Office of the DNI, stated at an information technology security conference that there are 12 objectives supporting the Initiative's goal of comprehensively addressing the nation's cyber security concerns. They are:

1. Move towards managing a single federal enterprise network;
2. Deploy intrinsic detection systems;
3. Develop and deploy intrusion prevention tools;
4. Review and potentially redirect research and funding;
5. Connect current government cyber operations centers;
6. Develop a government-wide cyber intelligence plan;
7. Increase the security of classified networks;
8. Expand cyber education;
9. Define enduring leap-ahead technologies;
10. Define enduring deterrent technologies and programs;

11. Develop multi-pronged approaches to supply chain risk management; and
12. Define the role of cyber security in private sector domains [27]

One question often raised is whether the CNCI objectives are being pursued concurrently. Some security observers are concerned that the government's focus to date has been on securing federal security systems at the expense of other networks that have similar vulnerabilities. The disruption, or perceived accessing or manipulating of data in non-federal networks that contain personal financial information or manage the control systems of the nation's critical infrastructure could have significant economic, safety, and confidence-in-government implications. It is often noted that in the homeland security and law enforcement communities, where a great deal of post- 9/11 emphasis is placed on continuous information exchange and collaboration, efforts to secure the federal technology systems, while relegating state, local, and private sector organizations to lower standards of security, will simply redirect or delay risk that inevitably accompanies increased collaboration. This concern is often expressed by non-federal governmental entities which rely on and routinely coordinate efforts with the U.S. government but have not been apprised of the plans or resources accompanying the CNCI.

Given the secretive nature of the CNCI, one of the common concerns voiced by many security experts is the extent to which non-federal entities should have a role in understanding the threat to the nation's teleco-mmunications and cyber infrastructure and assist with providing advice, assistance, and coordination in preparation and response for ongoing and future intrusions and attacks [28]. As telecommunications providers and internet service providers are corporate entities residing in the private sector, and are relied upon heavily to support federal government activities and services, many cyber-security observers suggest that a comprehensive approach to an effective monitoring, defending, and responding regime is not possible without the collaboration and expertise of the nation's cyber sector owners and operators. As evidenced in the twelve objectives of CNCI, it appears the federal government focus is on the prevention aspects of addressing potential threats to the nation's cyber and telecommunications infrastructure. In contrast, the primary response and recovery activities associated with previous network breaches have been addressed by the private sector entity that has been the victim of the attack. In an apparent admission of the need for further transparency and enhanced public-private partnership to better fulfill the goals of the CNCI, former President Bush's Assistant

Secretary of Cybersecurity and Telecommunications at the Department of Homeland Security (DHS), Greg Garcia, recently stated that "there was too much classified (about the CNCI) which was not helpful politically and not helpful in getting the word out." Acknowledging the balance between incorporating the view of non-federal entities and the concern of allowing those that wish to use cyber activities to cause harm, Assistant Secretary Garcia went on to further state that the Department had to "walk the line between raised awareness of what was being accomplished and not letting out too much information that could cause us to be targeted. Still, too much was kept secret" [29].

Based on the number of unknowns concerning the CNCI and the apparent lack of inclusiveness with the private sector telecommunication and internet providers, some analysts are concerned that future opportunities for successfully ascertaining known and future threats and developing a comprehensive set of legal and policy responses may not be achievable. An apparent Obama Administration goal for the current 60-day cyber security review is a more transparent and coordinated approach to the nation's cyber security risks with the perceived end result being that all affected parties are consulted and given the opportunity to provide advice and assistance in proposing changes to existing legislation, policy, and processes [30].

Legal Authorities for Executive Branch Responses to Cyber Threats

As discussed, the CSIS report on Securing Cyberspace for the 44th Presidency recommends executive action to protect U.S. cyberspace.31 This and other calls for executive action, together with the 60-day review of the CNCI, implicate questions regarding legal authorities and the appropriate roles of the two political branches in the cybersecurity context. Questions concern the adequacy of existing statutes and the potential need for new legislation to address the modern threat. In addition, for actions not authorized by the existing statutory framework, questions arise regarding the extent of inherent authority for executive-branch responses under the U.S. Constitution.

To be legally authorized, the CNCI and any other executive-branch action must have some basis in statutory or constitutional law [32] Several disparate legal authorities offer potential bases for executive responses to cyber threats. These include: (1) various provisions in the criminal code that establish federal cybercrime offenses and authorize prosecution; (2) statutes, such as the

Federal Information Security Management Act, [33] which direct executive agencies to establish specific administrative procedures to prevent cyber attacks; (3) more general statutes authorizing executive management of federal agencies; (4) the Authorization for Use of Military Force passed by Congress in 2001, [34] which empowered the President to use "all necessary and appropriate" force against perpetrators of the 9/11 terrorist attacks or those who harbor them; and (4) executive powers inherent in the Commander-in-Chief clause or other constitutional provisions.

Because the CNCI objectives appear to include broad governmental reforms and enhanced partnerships with the private sector, at least some actions contemplated by the CNCI likely fall outside of the relatively straightforward and narrow delegations of authority granted by statutes that specifically address cybersecurity, such as federal criminal law provisions and the Federal Information Security Management Act. As previously noted, the Federal Information Security Management Act requires federal agencies to take steps, such as establishing a Chief Information Officer position, to protect their computer systems from cyber intrusions [35]. In the criminal law context, the federal computer fraud and abuse statute outlaws intrusions upon the security of government computer systems, and in some cases upon the security of computers used in interstate commerce, by trespassing, threats, damage, espionage, or corrupt use of government computers as instruments of fraud [36]. It is likely that some cybersecurity measures envisioned by the CNCI objectives fall outside the scope of both statutory schemes. Most criminal provisions are reactive by nature; they generally do not authorize preventative measures to defend against potential cyber threats, and jurisdictional and practical hurdles could hamper law enforcement's authority over a computer hacker operating abroad. In contrast, the Federal Information Security Management Act and related statutes, like the CNCI, take a preventative approach to stopping cyber intrusions. However, they require federal agencies to take administrative measures that are relatively modest compared with the objectives of the CNCI.

It is possible that some measures contemplated by the CNCI would find authority in statutes that do not explicitly address cyber threats. For example, statutes authorizing executive management of the civil service might authorize some changes to government internet portals or changes in agency personnel [37]. However, such statutes do not address cybersecurity explicitly, nor do they authorize actions taken outside the realm of administrative measures in federal agencies.

Therefore, the existing statutory framework may not provide adequate authority for at least some responses contemplated by CNCI objectives. To fill that possible gap, or to adopt alternative or supplemental approaches, Congress may determine that new legislation is appropriate. Potential legislative approaches are discussed infra [38]. However, even if current statutory law is inadequate to protect the country against cyber attacks, it is not necessarily inadequate in the sense of providing insufficient legal authority for the CNCI, because inherent constitutional powers provide an alternative source of legal authority for some executive branch actions. Thus, Congress could decline to act legislatively in some areas, perhaps choosing instead to work with the executive branch in a cooperative or oversight role. If it did so, the executive branch could act in a number of situations by relying on inherent powers under Article II of the U.S. Constitution or, in very limited circumstances, on the 2001 Authorization to Use Military Force [39].

The Supreme Court's separation-of-powers jurisprudence makes clear that the President may occasionally act pursuant to his inherent powers under the Constitution without express or implied authorization from Congress [40]. Powers most relevant to the CNCI include the President's war and foreign affairs powers.

Separation of Powers in National Security Matters

The Constitution divides powers relating to national security between the executive and legislative branches. Article I of the U.S. Constitution empowers Congress to "declare war," "raise and support armies," "provide and maintain a navy," and "make rules for the government and regulation of the land and naval forces." [41] Article II states that the "President shall be Commander in Chief of the Army and Navy of the United States, and of the Militia of the several States"[42]. As a preliminary matter, invocation of war powers begs a question regarding the scope of the Commander in Chief's role in a modern conflict that, not least in the context of cyber warfare, defies traditional military strategies. Many facets of the CNCI – such as components directing planning, development, and education – fall outside of traditional definitions of war. In addition, war powers would likely not apply to actions which mandate private sector security measures. However, many believe the Commander in Chief power extends beyond warfare to encompass a broad conception of national security. In addition, although the phrase "war powers" evokes international conflicts, it seems that the President's war powers authorize at least some domestic action. For example, some have argued that the President's Commander in Chief power authorizes him to create a domestic intelligence agency [43].

Alternatively, the President's foreign affairs powers might provide an inherent constitutional authorization for executive action on cybersecurity. Given modern communications technology and the ease of travel, it is increasingly difficult to draw clean lines between foreign and domestic affairs. Congress' attempts to distinguish between foreign and domestic actors in other areas impacted by rapidly changing technological environments serve as examples. For instance, in the context of electronic surveillance, statutory provisions have progressed from drawing definitive distinctions between people located in the United States versus abroad in the original Foreign Intelligence Surveillance Act to a 2007 amendment excluding from the scope of foreign surveillance any person "reasonably believed" to be located abroad [44].

Finally, the President might assert that his oath-based obligation to defend the nation from imminent threats, sometimes termed the "emergency theory," provides a constitutional basis for executive action to prevent cyber intrusions or attacks. Presidents have relied on this authority very rarely [45].

Assuming that the President's war or foreign affairs powers extend to national security efforts such as the CNCI, the next question is whether, and in what circumstances, the executive branch exercise of such powers might be constrained by congressional action. As discussed, Congress and the President share powers to address matters of national security, and no precise line divides the powers of the two political branches. Some have identified a narrow sphere of Article II authority, sometimes called "preclusive" power, [46] which congressional action cannot limit. For most situations, however, Justice Robert Jackson's concurring opinion in Youngstown Steel and Tube Co. [47] establishes the leading doctrine governing the executive's inherent constitutional authority vis-a-vis Congress [48]. Justice Jackson's three-category framework requires courts to evaluate, where possible, the interplay between congressional intent and executive action in the context of the Constitution's allocation of powers. This exercise is made more difficult by the murky nature of a small category of inherent constitutional powers some believe are reserved to the President alone.

During the Korean War, President Truman signed an executive order directing the Commerce Secretary to take control of the nation's steel mills in order to prevent a national steelworkers' strike. In Youngstown, also known as the "Steel Seizure Case," the government claimed that presidential powers inherent in Article II provisions, most notably the Commander-in-Chief power, authorized President Truman's action [49]. To prove this claim, the government characterized the industry seizure as an action of a Commander in Chief, prompted by exigencies of war: steel production was necessary for military operations in Korea [50]. The Supreme Court rejected this claim, [51] but justices reached the conclusion by different analytical routes.

Writing for the majority, Justice Black took the hard-line view that the Commander-in-Chief clause gives the President no substantive authority. He emphasized that controlling private property to affect labor disputes "is a job for the nation's lawmakers" [52].

In contrast, Justice Jackson argued that the President's inherent constitutional powers "fluctuate," from relatively high when authorized by Congress, to their "lowest ebb" when a president "takes measures incompatible with the express or implied will of Congress."[53] Specifically, Justice Jackson articulated three categories of executive action: (1) action supported by an express or implied grant of authority from Congress; (2) a "zone of twilight" between the other categories, in which "congressional inertia" can occasionally "enable, if not invite, measures on independent presidential responsibility"; and (3) action that conflicts with statutes or congressional intent [54]. Actions in the first category enjoy congressional support and thus might not need to rely solely on an inherent constitutional powers argument; assuming that Congress acted pursuant to an enumerated Article I power in delegating the authority, these actions are clearly authorized unless they violate another constitutional provision. Actions in the second, "zone of twilight" [55] category prompt a complicated, totality-of-the circumstances inquiry, in which courts determine congressional intent vis-a-vis executive action. Actions that fall within the third category – that is, actions that conflict with statutory law – generally lack constitutional authority, unless the action is one of the few types of actions over which the President has exclusive authority. In *Young*stown, Justice Jackson found that President Truman's actions fit within the third category, because Congress had not left the issue of property seizure during labor disputes to an "open field"; rather, Congress had passed statutes designed to stabilize markets when government required supplies [56]. On this basis, Justice Jackson joined the majority to strike down President Truman's seizure of the steel industry [57].

Given the existing statutory framework, at least some potential responses to cyber threats would likely fall outside of the first of Justice Jackson's categories. Congress has not expressly authorized the cybersecurity reforms proposed by the CNCI, nor do the Federal Information Security Management Act or related statutes appear to impliedly authorize all potential cybersecurity protections. In addition, although the use of cyber force might have congressional authorization under the 2001 Authorization for Use of Military Force [58] if directed against an al Qaeda or Taliban operative, the Supreme Court has appeared to foreclose reliance on the Authorization as a basis for any action that is not a "fundamental" incident to the use of force against those responsible for the 9/11 attacks. The 2001 joint resolution authorized the use of "all necessary and appropriate force against those nations, organizations, or persons he determines planned, authorized, committed, or aided" the 9/11 attacks [59]. In Hamdi v. Rumsfeld, the Supreme Court held that capture and detention of Taliban members constituted "so fundamental and accepted an incident to war as to be an exercise of the 'necessary and appropriate force' Congress has authorized the President to use."60 The Court seemed reluctant to interpret the Authorization as extending to detentions beyond this "limited category." [61] Cyber security efforts that focus on information gathering activities may parallel the role of intelligence collection as a "central component of the war on terrorism." [62] However, not all cybersecurity threats fit logically within the scope of the so-called War on Terror. Cyber intrusions conducted by individual computer hackers, not supported by or aligned with a nation or terrorist organization, are perhaps best characterized as ordinary criminal activity whereas orchestrated intrusions by foreign security or intelligence entities might belong in a category of routine foreign-intelligence gathering. Neither activity appears to fit the mold of wartime operations. On the other hand, to the extent that the primary aim of the War on Terror is to prevent terrorists from harming U.S. civilians or assets, one might argue that defending the United States against threats to the U.S. cyber and telecommunications infrastructure fits squarely within the War's parameters [63] Nonetheless, it seems unlikely that all aspects of the CNCI would fit within the Hamdi interpretation of the 2001 Authorization.

On the other hand, unless Congress takes legislative action that contravenes a proposed executive response, the third category in Justice Jackson's framework is inapplicable. In contrast to intelligence collection efforts through the use of electronic surveillance, which Congress explicitly limited in the Foreign Intelligence Surveillance Act,64 Congress has not expressly limited executive action on cybersecurity. Although Congress has not left the cybersecurity arena an entirely "open field," by virtue of its modest actions with regard to the Federal Information Security Management Act and related provisions, it has not occupied the field to the extent that it had occupied the arena of labor regulation at issue in Youngstown.

Therefore, the CNCI and other potential executive actions taken to address cybersecurity likely fall within Justice Jackson's second, "zone of twilight" category, in which the executive and legislative branches have shared authority to act. A 1981 case, Dames and Moore v. Regan, refined the Supreme Court's approach to evaluating actions that lie within this "zone of twilight" [65]. In Dames, then-Justice Renquist, writing for the majority, clarified that in "zone of twilight" cases, the analysis, at least so far as separation-of-powers principles are concerned, "hinges on a consideration of all the circumstances which might shed light on the views of the legislative branch toward [the executive's] action, including 'congressional inertia, indifference or quiescence.'"[66]. Thus, the inquiry in such cases becomes a balancing act, aimed toward ascertaining Congress' relationship to the subject matter at issue. In the context of the CNCI, Congress' actions to date on cybersecurity have been primarily criminal or administrative and do not represent a comprehensive response to the issue. In addition, the CNCI involves intelligence and foreign affairs issues that traditionally lie within the purview of the executive branch. Therefore, at least until Congress takes further action in the cybersecurity area, it appears that the executive branch is not precluded from implementing the CNCI or other cybersecurity responses under Justice Jackson's Youngstown framework.

A final issue is whether responses to cybersecurity intrusions or attacks might be part of the narrow realm of "preclusive" constitutional powers belonging to the President [67]. Although the scope of, and even the constitutional authority for, such powers has never been fully defined, scholars have noted that a few key rules form a "rarely questioned narrative" regarding arenas in which Congress traditionally defers to executive action [68] For example, traditional notions dictate executive direction of wartime campaigns [69]. Likewise, the Supreme Court has characterized the President as the "sole organ" of the country in conducting foreign affairs [70]. In addition, some have suggested a distinction between offensive utilization of cyber weapons versus defensive shield to stop attacks: [71] whereas the President must obtain congressional authorization before committing U.S. armed forces in an offensive action, the President's has the exclusive power to repel attacks made against the United States.

Despite this narrative, however, no definitive boundaries have been defined for any such preclusive powers. Perhaps for that reason, Justice Jackson made clear in his Youngstown concurrence that the realm of any such preclusive powers must be carefully scrutinized [72]. Thus, although many executive actions in the cyber area would likely fall within the scope of Article II powers for ensuring national security, most actions would probably falls outside of the narrow categories of exclusive executive authority to conduct wartime campaigns and international relations. Similarly, even if the President has an exclusive power to lead the military in defensive actions, actions might not be clearly enough a defensive response to a military threat to trigger an exclusive presidential power [73].

Thus, it appears that the Youngstown framework would apply to a review of the President's authority to implement responses such as the CNCI. Thus, if Congress passed conflicting legislation in the cybersecurity area, some executive actions could be constrained. Alternatively, congressional legislation granting explicit authority for cybersecurity measures would more firmly confirm the executive authority to act in that area.

It is possible that the Supreme Court will address the constitutional authorities for national security in a future case. Youngstown represents one of only a small number of cases in which the Supreme Court has reached questions regarding the political branches' shared powers under the Constitution. Modern threats might necessitate new definitions within the Court's separation-of powers jurisprudence. For example, as cyber activities and telecommunication architectures are networked globally, with it often being difficult to ascertain where an attack or intrusion emanates, distinctions based on notions of conventional war may seem increasingly inconsistent with the modern Commander-in-Chief role.

CONGRESSIONAL CONSTRAINTS ON EXECUTIVE ACTION

Even if the CNCI or future cybersecurity initiatives are grounded in statutory or constitutional authority, questions will nonetheless arise regarding the degree to which legislative oversight is appropriate. Congress has attempted to obligate the President to report to relevant congressional leaders for actions taken pursuant to war powers or as part of intelligence operations. In 1973, Congress passed the War Powers Resolution to "fulfill the intent of the framers of the Constitution of the United States and insure that the collective judgment of both the Congress and the President will apply to the introduction of United States Armed Forces into hostilities." [74] Although presidents since the Resolution's passage have maintained that the Resolution unconstitutionally limits presidential authority, presidents have in many cases submitted documents for Congress that are "consistent with" the Resolution's requirements [75]

Similarly, after the Iran-Contra affair, Congress passed legislation increasing congressional oversight of intelligence activities, including significant and anticipated intelligence activities, and covert actions [76]. To the extent consistent with due regard for preventing unauthorized disclosure of classified information regarding sensitive intelligence sources and methods, current law requires that congressional intelligence committees be kept fully informed regarding intelligence activities. If the President determines that it is essential to meet extraordinary circumstances affecting vital U.S. interests, a presidential finding regarding a covert action may be limited to a small number of congressional leaders [77].

With respect to the CNCI, a key question is whether ongoing or potential U.S. cyber activities, defensive and offensive, may fall within the sphere of a covert activity or an intelligence activity and thus trigger reporting requirements. The statutory definition of "covert actions" includes "activity or activities of the United States Government to influence political, economic, or military conditions abroad, where it is intended that the role of the United States Government will not be apparent or acknowledged publicly," but excludes activities conducted for the purpose of gathering intelligence and "traditional" diplomatic, military, or law enforcement activities [78]. The definition of "intelligence activity" is broader; it includes covert actions and "financial intelligence activities" [79]. Because the definitions focus on the influence, rather than the presence, of conditions abroad, it appears that cyber actions targeting or even defending against cyber threats, even if conducted inside the United States, could trigger reporting requirements.

In addition to the potential application of ongoing reporting requirements, Congress could elicit information regarding executive actions by virtue of its enumerated power to control spending. The 110th Congress took several steps to obtain information regarding the CNCI in that manner. A continuing resolution, passed by Congress and signed into law in September 2008, withholds \$127 million of a \$313.5 million appropriation for cybersecurity until House and Senate appropriations committees "receive and approve a plan for expenditure for [the CNCI] that describes the strategic context of the program; the specific goals and milestones set for the program; and the funds allocated to achieving each of those goals."[80]. In addition, the Senate Committee on Homeland Security and Governmental Affairs held a closed hearing in March 2008 regarding the CNCI and later obtained answers to some questions regarding the initiative [81]. Finally, as part of a larger Homeland

Security Authorization bill, S. 3623, Senator Lieberman introduced legislation during the 110th Congress that would provide for congressional oversight of the CNCI and establish "a robust National Cyber Security Center with the mission of coordinating and enhancing federal efforts to protect government networks."[82] As an authorization bill for the DHS has not been passed since the creation of the Department, whether the proposed legislative oversight efforts will be effective remains to be seen. Also, as with many programs associated with intelligence community activities and defense, concerns regarding committee jurisdiction in the areas of oversight, authorization, and appropriations might be raised for the CNCI.

Policy Considerations and Congressional Options

As with executive control over covert actions, foreign affairs, and intelligence gathering, strong justifications support the assertion that the executive branch is best suited to take reasonable and necessary actions to defend the country against cyber-based threats. One such justification stems from the broad diversity of cybersecurity threats: the President is arguably best positioned to take a leadership role or create a uniform response to span the range of cyber vulnerabilities. In addition, the executive branch is likely most able to integrate intelligence-gathering, military, and other vehicles for addressing the cybersecurity challenge. However, in order for Congress to maintain ongoing awareness of CNCI plans and activities and to effectively perform its constitutional duties of oversight based on a thorough understanding of executive branch activities, some security experts suggest a range of legislative activities that might be required. Congress might choose to:

- determine the most appropriate and effective organizational entity in which the nation's principal cybersecurity prevention, response, and recovery responsibilities should reside; [83]
- require the senior U.S. government official in charge of all CNCI related activities be a Senate confirmable position to facilitate ongoing information exchange regarding Initiative plans and areas of progress and difficulty;
- enact legislative language recognizing and defining the classified and unclassified aspects of the CNCI and the need for greater transparency and inclusiveness;

- require the new Administration to develop and revise annually a classified and unclassified national cyber security strategy and intelligence community generated National Intelligence Estimate that provides Congress, the telecommunications industry, and the American public information related to the CNCI, the current and strategic cyber threats facing the nation, and programs being implemented to prepare for evolving technological risks;
- define the privacy and civil liberty considerations that should accompany all aspects of the CNCI;
- include legislative language in applicable authorizations bills to establish a programmatic foundation for CNCI related programs and suggest funding for current and future year's activities; or
- identify and codify relevant laws defining a national security related cyber offense against the United States, offensive versus defensive cyber activities, and
- the situations in which the Congress should be notified prior to the United States undertaking an offensive or counteroffensive cyber act.

CONCLUSION

As discussed, multiple policy considerations, including the novel and dispersed nature of cyber threats, might justify an executive-led response to cybersecurity. In response to calls for executive action, questions have arisen regarding the adequacy of legal authorities justifying executive responses to cyber threats. Although existing statutes might support some executive actions, the current statutory framework likely does not address all potential actions. Thus, the extent of inherent powers under Article II of the Constitution and the appropriate roles of the two political branches in this emerging national security arena are relevant considerations. Arguably, both the statutory framework and separation of powers analyses might need to be modernized to address appropriate roles in protecting the United States against modern cyber threats.

Finally, even if executive branch responses are authorized, Congress retains an oversight role vis- à-vis the CNCI or other presidential initiatives, for several reasons. First, if Congress passed statutes in contravention of the President's efforts, the President's authority to proceed with those efforts would become more questionable under the Youngstown framework. Second, as with covert actions, Congress likely has a legislative oversight role, even if that role merely requires notice of significant actions. Finally, Congress could ultimately withhold funding for the CNCI or specific aspects of the program should it not receive the necessary information to make an assessment of the activities related to each of the twelve objectives.

REFERENCES

[1] Center for Strategic and International Studies, Securing Cyberspace for the 44th Presidency: A Report of the CSIS Commission on Cybersecurity for the 44th Presidency (2008).

[2] Department of Homeland Security, Fact Sheet: DHS 2008 End of Year Accomplishments (Dec. 18, 2008), http://www.dhs.gov/xnews/releases /pr_1229609413187.shtm.

[3] *Id.*

[4] See, e.g., House Permanent Select Committee on Intelligence, Cyber Security: Hearing on the Nation's Cyber Security Risks, 110th Cong. (Sept. 18, 2008); House Homeland Security Committee, Cybersecurity Recommendations for the Next Administration: Hearing Before the Subcommittee on Emerging Threats, Cybersecurity and Science and Technology, 110th Cong. (Sept. 16, 2008).

[5] July 17, 2008 speech at Purdue University. As of the date of this report a national Cyber Security Advisor has not been named.

[6] Center for Strategic and International Studies, *Securing Cyberspace for the 44th Presidency: A Report of the CSIS Commission on Cybersecurity for the 44th Presidency* 3 (2008) (including "do not start over" as one of its recommendations for the 44th presidency).

[7] The White House, Office of the press Secretary, *President Obama Directs the National Security and Homeland Security Advisors to Conduct Immediate Cyber Security Review* (Feb. 9, 2009), http://www.whitehouse.gov/the_press_office/AdvisorsToConductImmediateCyberSecurityReview/.

[8] Department of Defense, Defense Science Board, *Defense Imperatives for the New Administration* 3 (2008), http://www.acq.osd.mil/dsb/reports/ 2008-11-Defense_Imperatives.pdf.

[9] Peter Eisler, Reported Raids on Federal Computer Data Soar, USA Today (Feb. 17, 2009), http://www.usatoday.com/news/washington/ 2009-02-16-cyber-attacks_N.htmcsp=34. Based on data reportedly provided to USA Today, the U.S. Computer Emergency Readiness Team

(US-CERT), a Department of Homeland Security entity, found that known cyberattacks on U.S. government networks rose 40% in 2008 compared to 2007. While this survey focused on U.S. government computer systems, telecommunications networks are maintained by private industry, and any degradation to these services or components would necessarily have negative implications for both public and private cyber activities.

[10] For more information on cyberattackers' capabilities, see CRS Report RL33123, *Terrorist Capabilities for Cyberattack: Overview and Policy Issues*, by John Rollins and Clay Wilson.

[11] 42 U.S.C . §5195c(e). For more on U.S. efforts to protect critical infrastructures, see CRS Report RL30153, Critical Infrastructures: Background, Policy, and Implementation, by John D. Moteff.

[12] Of note, many of the cyber-related incidences that were found to have negatively affected control systems connected to physical infrastructure components were resolved as being the work of current or former employees who had access to and knowledge of the architecture of the affected network.

[13] Jeanne Meserve, *Staged Cyber Attack Reveals Vulnerability in Power Grid*, CNN online (Sep. 26, 2007), http://www.cnn.com/2007/US/09/26/power.at.risk/index.html#cnnSTCVideo. A video of the experiment, named Project Aurora, and the resulting damage to the generator is available on the CNN website.

[14] *See* Center for Strategic and International Studies, *Securing Cyberspace for the 44th Presidency: A Report of the CSIS Commission on Cybersecurity for the 44th Presidency* 12 (2008) ("we expected damage from cyber attacks to be physical (opened floodgates, crashing airplanes) when it was actually informational").

[15] House Permanent Select Committee on Intelligence, *Cyber Security: Hearing on the Nation's Cyber Security Risks*, 110th Cong. (Sept. 18, 2008) (statement of Paul Kurtz, Former Senior Director, Critical Infrastructure Protection, White House Homeland Security Council).

[16] The Charlie Rose Show, "Interview of Mr. Mike McConnell, Director of National Intelligence," PBS, January 8, 2009.

[17] U.S. Congress, Senate Select Committee on Intelligence, Annual Threat Assessment of the Intelligence Community: Hearing on the Threats to the Nation, 111th Cong. (Feb. 12, 2009).

[18] Melissa Hathaway, Cyber Security – An Economic and National Security Crisis, Intelligencer: Journal of U.S. Intelligence Studies, Fall 2008 at 31-6.

[19] U.S. Congress, Senate Select Committee on Intelligence, Annual Threat Assessment of the Intelligence Community: Hearing on the Threats to the Nation, 111th Cong. (Feb. 12, 2009).

[20] In November, 2008, it was reported that the Department of Defense notified all organizations to stop using portable storage devices as it has become "apparent that over time, our posture to protect networks and associated information infrastructure has not kept pace with adversary efforts to penetrate, disrupt, interrupt, exploit or destroy critical elements of the global information grid." Noah Shachtman, Military USB Ban Meant to Stop Adversary Attacks, Wired Blog Network (Nov. 20, 2008), http://blog.wired.com/defense/2008/11/military-usb-ba.html. Also, it has recently been reported that some U.S. military units have resorted to disconnecting computer networks from the internet for fear of cyber related risks and a concern that the affected organization may not be managing its network properly thus "making everyone else vulnerable" to an attack. Noah Shachtman, *Air Force Unplugs Bases' Internet Connections*, Wired Blog Network (Feb. 18, 2000), http://b log.wired.com/defense/2009/02/air-force-cuts.html.

[21] See Center for Strategic and International Studies, Securing Cyberspace for the 44th Presidency: A Report of the CSIS Commission on Cybersecurity for the 44th Presidency 12 (2008) at 67.

[22] *See, e.g., Id.* at 69 (stating that the Act "has become a paperwork exercise rather than an effective measure of network security"). The Federal Information Security Management Act is Title III of the E-Government Act of 2002, P.L. 107-347, 116 Stat. 2899 (codified at 44 U.S.C. §3541 *et. seq.*). Among other things, it created a position of Chief Information Officer within each federal agency.

[23] Federal Information Security Management Act of 2008, S. 3474, 110th Cong. (2008). The bill was favorably reported by the Senate Homeland Security and Government Affairs Committee and was placed on the Senate calendar. It has not yet been reintroduced during the 111t[h] Congress.

[24] 44 U.S.C. §3506 (requiring Chief Information Officer positions). The Clinger-Cohen Act is the name given to the Federal Acquisition Reform Act of 1996 and the Information Technology Management Reform Act of 1996, which passed as Sections D and E, respectively, of the National

Defense Authorization Act for Fiscal Year 1996, P.L. 104- 106, 110 Stat. 642, 679 (1996).

[25] House Homeland Sec. Comm., Cybersecurity Recommendations for the Next Administration: Hearing Before the Subcommittee on Emerging Threats, Cybersecurity and Science and Technology, 110th Cong. (Sept. 16, 2008) (statement of James A. Lewis, Director and Senior Fellow, Center for Strategic and International Studies).

[26] For more information on presidential directives, see CRS Report 98-611, Presidential *Directives: Background and Overview*, by Harold C. Relyea/

[27] Wyatt Kash, Government Computer News, *Details Merge About the President's Cyber Plan* (Nov. 21, 2008), http://gcn.com/ Articles /2008/11/21/Details-emerge-about-Presidents-Cyber-Plan.aspx?Page=4.

[28] It is unknown whether non-federal entities have been invited to participate in the previously mentioned President's 60-day cyber security review that commenced on February 9, 2009.

[29] Jill Aitoro, *Bush's Cyber Chief Calls National Security Initiative Too Secret*, Nextgov (Feb. 11, 2009), http://www.nextgov.com/nextgov/ ng_200 90211_6858.php.

[30] See Press Release, White House, *President Obama Directs the National Security and Homeland Security Advisors to Conduct Immediate Cyber Security Review*, (Feb. 9, 2009), http://www.whitehouse.gov/ the_pres s_office/AdvisorsToConductImmediateCyberSecurityReview/.

[31] U.S. Department of Homeland Security, DHS Data Privacy and Integrity Advisory Committee, *Letter to the Secretary Regarding Data Privacy and Integrity Recommendations*, Executive Summary, Feb. 5, 2009, p. 4.; Center for Strategic and International Studies, *Securing Cyberspace for the 44th Presidency: A Report of the CSIS Commission on Cybersecurity for the 44th Presidency.*

[32] Because the federal government is a government of limited powers, executive actions must find support in either: (1) a power enumerated under Article II of the U.S. Constitution; or (2) authority delegated to the executive by Congress pursuant to one or more of Congress' enumerated Article I powers. Within this framework, some actions are impliedly authorized as means to achieve ends authorized by enumerated powers. See McCulloch v. Maryland, 17 U.S. 316 (1819) (upholding Congress' creation of a National Bank as a constitutionally valid means by which to exercise enumerated Article I powers).

[33] 44 U.S.C. §3541 *et. seq.*

[34] Authorization for Use of Military Force, P.L. 107-40, 115 Stat. 224 (2001). For background information on authorizations for use of military force and differences between such authorizations and declarations of war, see CRS Report RL31133, *Declarations of War and Authorizations for the Use of Military Force: Historical Background and Legal Implications*, by Jennifer K. Elsea and Richard F. Grimmett.

[35] 44 U.S.C. §3541 *et. seq.*

[36] 18 U.S.C. §1030. For an overview of federal cybercrime provisions, see CRS Report 97-1025, *Cybercrime: An Overview of the Federal Computer Fraud and Abuse Statute and Related Federal Criminal Laws*, by Charles Doyle.

[37] Statutes authorizing executive management of the civil service are codified in Title 5 of the U.S. Code.

[38] The extent of any new law would be limited by individual constitutional rights and by international laws of war.

[39] If the President has authority to act pursuant to powers inherent in the U.S. Constitution, then authority under the Authorization to Use Military Force is unnecessary, and visa versa. Under either source, the scope of executive power might depend upon the intent of and actions taken by Congress.

[40] The executive and legislative branches typically resolve disputes regarding the extent of executive authority without involving the courts. However, the Supreme Court is the final arbiter in such disputes. *See* David J. Barron and Martin S. Lederman, *The Commander in Chief at the Lowest Ebb – Framing the Problem, Doctrine, and Original Understanding*, 121 Harv. L. Rev. 689, 722-237 (2008).

[41] U.S. Const. Art. I, §8.

[42] U.S. Const. Art. II, §2, cl.1.

[43] RAND Corp., The Challenge of Domestic Intelligence in a Free Society: A Multidisciplinary Look at the Creation of a U.S. Domestic Counterterrorism Intelligence Agency 108 (2009) (arguing that for establishing a domestic intelligence agency, the Constitution "tilts the balance of power toward the President by virtue of the Commander-in-Chief clause").

[44] The Foreign Intelligence Surveillance Act of 1978, P.L. 95-511, 92 Stat. 1783 (1978) (codified as amended at 50 U.S.C. §§1801 et seq.); see also Protect America Act, P.L. 110-55 (2007).

[45] Some attorneys within the Bush Administration relied on the emergency powers argument to assert that President Bush had inherent authority to use military force in the war on terror. *See, e.g.*, Memorandum Opinion for the Deputy Counsel to the President, *The President's Constitutional Authority to Conduct Military Operations Against Terrorists and Nations Supporting Them* (Sept. 25, 2001), http://www.usdoj.gov/olc/warpowers925.htm.

[46] The term "preclusive" appeared in Justice Jackson's concurring opinion in *Youngstown Steel and Tube Co.*, 343 U.S. 579 (1952), when he referred to Article I authorities that, if exercised, would preclude a conflicting action by Congress as "at once so conclusive and preclusive [that they] must be scrutinized with caution." 343 U.S. at 638 (Jackson, J., concurring).

[47] 343 U.S. 579 (1952).

[48] *See* Hamdan v. Rumsfeld, 548 U.S. 557, 638 (2006) ("The proper framework for assessing whether executive actions are authorized is the three-part scheme used by Justice Jackson in his opinion in *Youngstown*").

[49] 343 U.S. at 587.

[50] *Id.*

[51] *Id.* The Court noted that "'theater of war' [is] an expanding concept." Id. Nonetheless, the Court "[could not] with faithfulness to our constitutional system hold that the Commander in Chief of the armed forces has the ultimate power as such to take possession of private property in order to keep labor disputes from stopping production." *Id.*

[52] *Id.*

[53] *Id.* at 635-38 (Jackson, J., concurring)

[54] *Id.*

[55] The phrase "zone of twilight" refers to the mesopelagic region of the ocean – the last region which light reaches, but it also has a non-scientific definition of an indefinite area between two conditions. Under Justice Jackson's framework, the President and Congress might have concurrent authority in this category, such that it is not always clear what, if any, power one branch has to supersede actions of the other.

[56] *Id.* at 639 (Jackson, J., concurring).

[57] *Id.*

[58] P.L. 107-40, 115 Stat. 224 (2001).

[59] P.L. 107-40, 115 Stat. 224 (2001).

[60] 542 U.S. 507, 518 (2004). However, the *Hamdi* court held that such authority is limited by detainees' rights under the due process clause. *Id.*

[61] *Id.*

[62] David J. Barron and Martin S. Lederman, *The Commander in Chief at the Lowest Ebb – Framing the Problem, Doctrine, and Original Understanding*, 121 Harv. L. Rev. 689, 714 (2008) ("a central component of the war against terrorism is, by its nature, the collection of intelligence").

[63] *See Id.* (noting that the war on terrorism differs from conventional conflicts, in part, because "the Executive has identified its principal goal in this conflict not as defeating the enemy in battle, but as preventing the enemy from 'fighting' in the first place").

[64] 50 U.S.C. §§1801 *et seq.*

[65] 453 U.S. 654 (1981).

[66] *Id.* at 669.

[67] Scholars have expressed doubts regarding the framers' intent to imbue the President with "preclusive" constitutional powers but nonetheless have argued that long-standing assumptions that such powers exist have solidified their constitutional standing. See, e.g., David J. Barron and Martin S. Lederman, *The Commander in Chief at the Lowest Ebb – Framing the Problem, Doctrine, and Original Understanding*, 121 Harv. L. Rev. 689, 802 (2008).

[68] *See, e.g Id.* at 698. For more information regarding divisions between Congress' and the President's war powers and an analysis of that division in the context of the President's authority to use commit armed forces in Iraq, see CRS Report RL33 837, *Congressional Authority to Limit U.S. Military Operations in Iraq*, by Jennifer K. Elsea, Michael John Garcia, and Thomas J. Nicola.

[69] *See Hamdan v. Rumsfeld*, 548 U.S. 557, 591-92 (2006) (citing *Ex Parte Milligan*, 71 U.S. 2, 139-40 (1866)). *But see* War Powers Resolution, 50 U.S.C. §§1541-1548, discussed *infra*.

[70] *See United States v. Curtiss-Wright Export Co.*, 299 U.S. 304, 319 (1936) ("'The President is the sole organ of the nation in its external relations, and its sole representative with foreign nations.'" (citing Annals, 6th Cong., col. 613 (statement of John Marshall))). However, the *Curtiss-Wright* case involved executive action that fell in the first of Justice Jackson's *Youngstown* categories – i.e., where Congress and the President acted in concert. Thus, although the case has helped to form a narrative regarding executive-branch prerogative in international

relations and has occasionally been cited to support the proposition that the President has some preclusive foreign affairs powers under the Constitution, it would misstate the *Curtiss-Wright* holding to assume that it recognized any broad preclusive foreign relations power.

[71] Aside from the operational distinction that may be made with respect to the types of cyber activities the U.S. may undertake, the offensive versus defensive distinction may also be worth considering from an organizational perspective. Agencies responsible for protecting the government's websites and launching counter-offensive attacks may not be the same entities responsible for assisting in the recovery phase of an attack of national security significance on a U.S. cyber or telecommunications hosted network.

[72] 343 U.S. at 638 (Jackson, J., concurring).

[73] In the context of modern national security threats, the line between offensive and defensive action is not easily deciphered. For example, the United States captured and detained a large number of alleged enemy combatants in the course of post-September 1 1th military operations. Is the ongoing detention of such persons, often referred to as "preventative detention," an offensive action? The Supreme Court has upheld executive authority for such detentions on statutory rather than constitutional grounds; it has not addressed offensive versus defensive distinction. Hamdi, 542 U.S. 507. Thus, even if some components of the CNCI qualify as war-related activity, perhaps because they target cyber terrorists, little guidance exists regarding which actions might qualify as defensive rather than offensive actions under the traditional war powers analysis.

[74] War Powers Resolution, P.L. 93-148, 87 Stat. 555 (1973) (codified at 50 U.S.C. §§1541-1548); 50 U.S.C. §1541(a).

[75] For information Presidential actions vis-a-vis the War Powers Resolution, see CRS Report RL33532, *War Powers Resolution: Preside-ntial Compliance*, by Richard F. Grimmett.

[76] Fiscal Year 1991 Intelligence Authorization Act, P.L. 102-88, 105 Stat. 429 (1991) (codified as amended at 50 U.S.C. §§413, 413a, 413b)

[77] For more information on congressional oversight of covert actions, see CRS Report RL33715, *Covert Action: Legislative Background and Possible Policy Questions*, by Alfred Cumming.

[78] 50 U.S.C. §413b(e).

[79] 50 U.S.C. §413(f).

[80] Consolidated Security, Disaster Assistance, and Continuing Appropriations Act of 2009, P.L. 110-329, (2008).

[81] NSPD-54/HSPD-23 and the Comprehensive National Cyber Security Initiative: Hearing Before the Sen. Homeland Security and Governmental Affairs Comm., 110th Cong. (March 4, 2008).

[82] S. 3623, 110th Cong. §§601-08 (2008); 154 Cong. Rec. S9687 (daily ed. Sept. 26, 2008) (statement of Sen. Lieberman).

[83] Possible organizational constructs for such an entity range from a single entity placed in charge of all phases of U.S. cyber activity to a coordination office with the authority and responsibility to compel other organizations to adhere to the President's cyber strategy. Entities often noted as having a significant contribution to the U.S. cyber activity, which could add capability and resources to the CNCI's capabilities, include the cyber and telecommunications industries, intelligence and law enforcement communities, and academia.

In: Computer Networks
Editor: Raymond S. Grigello

ISBN: 978-1-61209-596-7

Chapter 5

SPYWARE: BACKGROUND AND POLICY ISSUES FOR CONGRESS*

Patricia Moloney Figliola

ABSTRACT

The term "spyware" generally refers to any software that is downloaded onto a computer without the owner's or user's knowledge. Spyware may collect information about a computer user's activities and transmit that information to someone else. It may change computer settings, or cause "pop-up" advertisements to appear (in that context, it is called "adware"). Spyware may redirect a Web browser to a site different from what the user intended to visit, or change the user's home page. A type of spyware called "keylogging" software records individual keystrokes, even if the author modifies or deletes what was written, or if the characters do not appear on the monitor. Thus, passwords, credit card numbers, and other personally identifiable information may be captured and relayed to unauthorized recipients.

Some of these software programs have legitimate applications the computer user wants. They obtain the moniker "spyware" when they are installed surreptitiously, or perform additional functions of which the user is unaware. Users typically do not realize that spyware is on their computer. They may have unknowingly downloaded it from the Internet by clicking within a website, or it might have been included in an

* This is an edited, excerpted and augmented edition of a CRS Report, Report RL32706 dated September 26, 2007.

attachment to an electronic mail message (e-mail) or embedded in other software.

The Federal Trade Commission (FTC) issued a consumer alert on spyware in October 2004. It provided a list of warning signs that might indicate that a computer is infected with spyware, and advice on what to do if it is.

Several states have passed spyware laws, but there is no specific federal law. Thus far, two bills have been introduced in the House of Representatives (H.R. 964 and H.R. 1525) and one has been introduced in the Senate (S. 1625). Both of the House bills have been reported and referred to the Senate. The Senate bill was referred to committee and no further action has been taken.

Note: This report was originally written by Marcia S. Smith; the author acknowledges her contribution to CRS coverage of this issue area.

BACKGROUND

Congress is debating whether to enact new legislation to deal with the growing problem of "spyware." Spyware is not well defined, but generally includes software placed on a computer without the user's knowledge that takes control of the computer away from the user, such as by redirecting the computer to unintended websites, causing "pop-up" advertisements to appear, or collecting information and transmitting it to another person. The lack of a firm definition of the term adds to the complexities of drafting new laws.

Opponents of new legislation argue that industry self-regulation and enforcement of existing laws are sufficient. They worry that further legislation could have unintended consequences that, for example, limit the development of new technologies that could have beneficial uses. Supporters of new legislation believe that current laws are inadequate, as evidenced by the growth in spyware incidents.

A June 2006 report on spyware enforcement by the Center for Democracy and Technology (CDT) summarizes active and resolved spyware cases at the FTC and the Department of Justice, and in individual states [1]

The main issue for Congress is whether to enact new legislation specifically addressing spyware, or to rely on industry self-regulation and enforcement actions by the FTC and the Department of Justice under existing law.

Advocates of legislation want specific laws to stop spyware. For example, they want software providers to be required to obtain the consent of an authorized user of a computer ("opt-in") before any software is downloaded onto that computer. Skeptics contend that spyware is difficult to define and consequently legislation could have unintended consequences, and that legislation is likely to be ineffective. One argument is that the "bad actors" are not likely to obey any opt-in requirement, but are difficult to locate and prosecute. Also, some are overseas and not subject to U.S. law. Other arguments are that one member of a household (a child, for example) might unwittingly opt-in to spyware that others in the family would know to decline, or that users might not read through a lengthy licensing agreement to ascertain precisely what they are accepting.

In many ways, the debate over how to cope with spyware parallels the controversy that led to unsolicited commercial electronic mail ("spam") legislation [2]. Whether to enact a new law, or rely on enforcement of existing law and industry self-regulation, were the cornerstones of that debate as well. Congress chose to pass the CAN-SPAM Act (P.L. 108-187). Questions remain about that law's effectiveness (see CRS Report RL31953). Such reports fuel the argument that spyware legislation similarly cannot stop the threat. In the case of spam, FTC officials emphasized that consumers should not expect any legislation to solve the spam problem — that consumer education and technological advancements also are needed. The same is true for spyware.

What is Spyware?

The term "spyware" is not well defined. Jerry Berman, President of CDT, explained in testimony to the Subcommittee on Communications of the Senate Commerce, Science, and Transportation Committee in March 2004 that "The term has been applied to software ranging from 'keystroke loggers' that capture every key typed on a particular computer; to advertising applications that track users' web browsing; to programs that hijack users' system settings." [3] He noted that what these various types of software programs "have in common is a lack of transparency and an absence of respect for users' ability to control their own computers and Internet connections." More recently, in June 2006, the Anti-Spyware Coalition (ASC) [4] issued a paper that defined spyware as "technologies deployed without appropriate user consent and/or implemented in ways that impair user control over:

- Material changes that affect their user experience, privacy, or system security;

- Use of their system resources, including what programs are installed on their computers; and/or
- Collection, use, and distribution of their personal or other sensitive information." [5]

Software programs that include spyware may be sold or available for free ("freeware"). They may be on a disk or other media, downloaded from the Internet, or downloaded when opening an attachment to an electronic mail (e-mail) message. Typically, users have no knowledge that spyware is on their computers. Because the spyware is resident on the computer's hard drive, it can generate pop-up ads, for example, even when the computer is not connected to the Internet.

One example of spyware is software products that include, as part of the software itself, a method by which information is collected about the use of the computer on which the software is installed, such as Web browsing habits. Some of these products may collect personally identifiable information (PII). When the computer is connected to the Internet, the software periodically relays the information back to another party, such as the software manufacturer or a marketing company. Another oft-cited example of spyware is "adware," which may cause advertisements to suddenly appear on the user's monitor — called "pop-up" ads. In some cases, the adware uses information that the software obtained by tracking a user's Web browsing habits to determine shopping preferences, for example. Some adware companies, however, insist that adware is not necessarily spyware, because the user may have permitted it to be downloaded onto the computer because it provides desirable benefits.

As Mr. Berman explained, spyware also can refer to "keylogging" software that records a person's keystrokes. All typed information thus can be obtained by another party, even if the author modifies or deletes what was written, or if the characters do not appear on the monitor (such as when entering a password). Commercial key logging software has been available for some time [6]. In the context of the spyware debate, the concern is that such software can record credit card numbers and other personally identifiable information that consumers type when using Internet-based shopping and financial services, and transmit that information to someone else. Thus it could contribute to identity theft [7]

Spyware remains difficult to define, however, in spite of the work done by groups such as the ASC and government agencies such as the Federal Trade Commission (FTC) [8]. As discussed below, this lack of agreement is often cited by opponents of legislation as a reason not to legislate. Opponents of anti-spyware legislation argue that without a widelyagreed-upon definition, legislation could have unintended consequences, banning current or future technologies and activities that, in fact, could be beneficial. Some of these software applications, including adware and keylogging software, do, in fact, have legitimate uses. The question is whether the user has given consent for it to be installed.

PREVALENCE OF SPYWARE

In October 2004, America Online (AOL) and the National Cyber Security Alliance (NCSA) [9] released the results of a survey of 329 dial-up and broadband computer users regarding online threats, including spyware [10]. According to the study:

- 80% of the computers they tested were infected with spyware or adware, and 89% of the users of those computers were unaware of it;
- the average infected computer had 93 spyware/adware components on it, and the most found on a single computer was 1,059; and
- most users do not recognize the symptoms of spyware — 63% of users with a pop-up blocker said they got pop-up ads anyway, 43% of users said their home page had been changed without their permission, and 40% said their search results are being redirected or changed.

Separately, Webroot Software, a provider of privacy and protection software, released the results of a survey of 287 corporate information technology managers on October 27, 2004. That survey concluded that although more than 70% of corporations expressed increased concern about spyware, less than 10% had implemented commercially available anti-spyware software [11].

FTC Advice to Consumers

The FTC has consumer information on spyware that includes a link to file a complaint with the commission through its "OnGuard Online website [12]. The FTC has also issued a consumer alert about spyware that lists warning signs that might indicate a computer is infected with spyware [13]. The FTC alert listed the following clues:

- a barrage of pop-up ads
- a hijacked browser — that is, a browser that takes you to sites other than those you type into the address box
- a sudden or repeated change in your computer's Internet home page
- new and unexpected toolbars
- new and unexpected icons on the system tray at the bottom of your computer screen
- keys that don't work (for example, the "Tab" key that might not work when you try to move to the next field in a Web form)
- random error messages
- sluggish or downright slow performance when opening programs or saving files.

The FTC alert also offered preventive actions consumers can take:

- update your operating system and Web browser software
- download free software only from sites you know and trust
- don't install any software without knowing exactly what it is
- minimize "drive-by" downloads by ensuring that your browser's security setting is high enough to detect unauthorized downloads
- don't click on any links within pop-up windows
- don't click on links in spam that claim to offer anti-spyware software
- install a personal firewall to stop uninvited users from accessing your computer.

Finally, the FTC alert advised consumers who think their computers are infected to get an anti-spyware program from a vendor they know and trust; set it to scan on a regular basis, at startup and at least once a week; and delete any software programs detected by the anti-spyware program that the consumer does not want.

Reviews of some of the commercially available anti-spyware programs are available in magazines such as PC World and Consumer Reports [http://www.pcworld.com/howto/article/0,aid,118215,00.asp] or at Spyware Warrior [http://www.spywarewarrior.com].

STATE LAWS

In March 2004, Utah became the first state to enact spyware legislation (although a preliminary injunction prevented it from taking effect, and the Utah legislature passed a new law in 2005 amending the 2004 act).[14]. In testimony to a House Energy and Commerce subcommittee in April 2004, then-FTC Commissioner Mozelle Thompson asked states to "be cautious" about passing such legislation because "a patchwork of differing and inconsistent state approaches might be confusing to industry and consumers alike." [15]

In 2006, at least 18 states have considered spyware legislation and at least three have enacted/adopted that legislation: Hawaii, Louisiana, and Tennessee. Detailed listings of spyware legislation from 2004, 2005, and 2006, are available on the National Council for State Legislature's website [16].

LEGISLATIVE ACTION – 110TH CONGRESS

During the 110th Congress, two bills have been introduced in the House of Representatives and one bill has been introduced in the Senate; the House has held two hearings.

H.R. 964 – Securely Protect Yourself against Cyber Trespass Act

The "SPY ACT" was introduced by Representative Towns on February8, 2007, and a hearing on it was held by the Committee on Energy and Commerce Subcommittee on Commerce, Trade and Consumer Protection on March 15, 2007 [17]. This bill would make it unlawful to engage in unfair or deceptive acts or practices to take unsolicited control of computer, modify computer settings, collect personally identifiable information, induce the owner or authorized user of the computer to disclose personally identifiable information, induce the unsolicited installation of computer software, and/or remove or disable a security, anti-spyware, or anti-virus technology. This bill would also require the FTC to submit two reports to Congress.

The first report would be on the use of cookies in the delivery or display of advertising; the second would be on the extent to which information collection programs were installed and in use at the time of enactment.

H.R. 964 was reported by the House Committee on Energy and Commerce on May 24, 2007, [18] and referred to the Senate Committee on Commerce, Science, and Transportation on June 7, 2007. No further action has been taken.

H.R. 1525 – Internet Spyware Prevention Act

The "I-SPY" Act was introduced by Representative Lofgren on March 14, 2007, and a hearing on it was held by the Committee on the Judiciary Subcommittee on Crime, Terrorism, and Homeland Security on May 1, 2007 [19]. This bill would amend the federal criminal code to impose a fine and/or prison term of up to five years for intentionally accessing a protected computer [20] without appropriate authorization by causing a computer program or code to be copied onto the protected computer and intentionally using that program or code in furtherance of another federal criminal offense. The bill would impose a fine and/or prison term of up to two years if the unauthorized access was for the purpose of—

- intentionally obtaining or transmitting personal information21 with intent to defraud or injure a person or cause damage to a protected computer
- intentionally impairing the security protection of a protected computer with the intent to defraud or injure a person or damage such computer.

H.R. 1525 was reported by House Committee on the Judiciary,where it was reported on May 21, 2007, [22] and then referred to the Senate Committee on the Judiciary on May 23, 2007. No further action has been taken.

H. Rep. 110-169. Available online at [http://www.congress.gov/cgi-lis/cpquery/ R?cp110:FLD010:@1(hr169)].

S. 1625 – Counter Spy Act

The Counter Spy Act was introduced by Senator Pryor on June14, 2007. This bill would prohibits unauthorized installation on a protected [23] computer of "software that takes control of the computer, modifies the computer's settings, or prevents the user's efforts to block installation of, disable, or uninstall software." It also would prohibit the installation of "software that collects sensitive personal information without first providing clear and conspicuous disclosure. . . and obtaining the user's consent. Additionally, S. 1625 would prohibit installation of software that "causes advertising windows to appear (popularly known as adware) unless: (1) the source is clear and instructions are provided for uninstalling the software; or (2) the advertisements are displayed only when the user uses the software author's or publisher's website or online service." This bill was referred to the Senate Committee on Commerce, Science, and Transportation on June 14, 2007. No further action has been taken.

APPENDIX

BILLS IN THE 108TH AND 109TH CONGRESSES

109th Congress

Two bills passed the House on May 23, 2005 — H.R. 29 (Bono) and H.R. 744 (Goodlatte) — both of which were very similar to legislation that passed the House in the 108th Congress.

Three bills were introduced in the Senate — S. 687 (Burns), which is similar to legislation that was considered in 2004, but did not reach the floor (S. 2145); S. 1004 (Allen); and S. 1608 (Smith). S. 687 and S. 1608 were ordered reported from the Senate Commerce Committee in 2005. At the markup that favorably reported S. 687, the committee rejected Senator Allen's attempt to substitute the language of his bill (S. 1004) for the text of S. 687. S. 687 was placed on the Senate Legislative Calendar under general Orders, Calendar no. 467, on June 12, 2006. S. 1608 was referred to the House Committee on Energy and Commerce Subcommittee on Commerce, Trade, and Consumer Protection, on April 19, 2006.

108th Congress

The House passed two spyware bills in the 108th Congress — H.R. 2929 and H.R. 4661. The Senate Commerce Committee reported S. 2145 (Burns), amended, December 9, 2004 (S.Rept. 108-424). None of these bills cleared that Congress.

The Senate Commerce, Science, and Transportation Committee's Subcommittee on Communications held a hearing on spyware on March 23, 2004. The House Energy and Commerce's Subcommittee on Telecommunications and the Internet held a hearing on April 29, 2004. The House passed two spyware bills (H.R. 2929 and H.R. 4661) and the Senate Commerce Committee reported S. 2145, but there was no further action.

References

[1] "Spyware Enforcement," CDT, June 2006, available online at [http://www.cdt.org/privacy/spyware/20060626spyware-enforcement.pdf].

[2] See CRS Report RL31953, "Spam": An Overview of Issues Concerning Commercial Electronic Mail, by Marcia S. Smith.

[3] Testimony to the Senate Committee on Commerce, Science, and Transportation, Subcommittee on Communications, March 23, 2004. Available on CDT's spyware site [http://www.cdt.org/privacy/spyware/] along with a November 2003 CDT report entitled Ghosts in Our Machines: Background and Policy Proposals on the "Spyware" Problem.

[4] The ASC is dedicated to building a consensus about definitions and best practices in the debate surrounding spyware and other potentially unwanted technologies. Composed of anti-spyware software companies, academics, and consumer groups, the ASC seeks to bring together a diverse array of perspective on the problem of controlling spyware and other potentially unwanted technologies. It's members include AOL, Cyber Security Industry Alliance, McAfee, Microsoft, SurfControl, US Coalition Against Unsolicited Commercial Email, and Yahoo. A complete list of the group's members is available online at [http://www.antispywarecoalition.org/about/index.htm].

[5] Anti-Spyware Coalition Definitions Document, June 2006, available online at [http://www.antispywarecoalition.org/documents/DefinitionsJune292006.htm].

[6] The existence of keylogging software was publicly highlighted in 2001 when the FBI, with a search warrant, installed such software on a suspect's computer, allowing them to obtain his password for an encryption program he used, and thereby evidence. Some privacy advocates argued that wiretapping authority should have been obtained, but the judge, after reviewing classified information about how the software works, ruled in favor of the FBI. Press reports also indicate that the FBI is developing a "Magic Lantern" program that performs a similar task, but can be installed on a subject's computer remotely by surreptitiously including it in an e-mail message, for example.

[7] For more on identity theft, see CRS Report RS22082, Identity Theft: The Internet Connection, by Marcia S. Smith; and CRS Report RL31919, Remedies Available to Victims of Identity Theft, by Angie A. Welborn.

[8] The FTC has a spyware information page on its website, [http://www.ftc.gov/spyware]. Further, a report from the FTC's April 2004 workshop on spyware is available online at [http://www.ftc.gov/os/2005/03/050307spywarerpt.pdf]. This report contains a discussion on the difficulties of defining spyware.

[9] According to its website [http://www.staysafeonline.org], NCSA is a public-private partnership, with government sponsors including the Department of Homeland Security and the FTC. Its Board of Officers includes representatives from Cisco Systems, Symantec, RSA Security, AOL, McAfee, Microsoft, and BellSouth.

[10] Largest In-Home Study of Home Computer Users Shows Major Online Threats, Perception Gap. Business Wire, October 25, 2004, 08:02 (via Factiva). The study is available online at [http://www.staysafeonline.info/news/safety_study_v04.pdf].

[11] Spyware Infiltration Rises in Corporate Networks, but Webroot Survey Finds Companies Still Neglect Threat. PR Newswire, October 27, 2004, 06:00 (via Factiva).

[12] Available at [http://onguardonline.gov/spyware.html].

[13] Available at [http://www.ftc.gov/bcp/conline/pubs/alerts/ spyware alrt.htm].

[14] WhenU, an adware company, filed suit against the Utah law on constitutional grounds. (WhenU's President and CEO, Avi Naider, testified to the Senate Commerce Committee's Subcommittee on Communications about spyware in March 2004. The Third Judicial District Court in Salt Lake City, Utah granted a preliminary injunction

on June 22, 2004, preventing the law from taking effect. See Judge Grants NY Pop-Up Company Preliminary Injunction Against Spyware Law. Associated Press, June 23, 2004, 06:06 (via Factiva).

[15] House Committee on Energy and Commerce. Hearing, April 29, 2004. Hearing transcript provided by the Federal Document Clearing House (via Factiva).

[16] See NCSL Electronic/Internet Privacy page ay [http://www .ncsl.org/programs/lis/privacy/ techprivacy.htm].

[17] Information on this hearing, including a list of witnesses, witness testimony, and a link to the hearing broadcast archive are available online at [http://energycommerce.house.gov/cmte_mtgs/110-ctcp_hrg. 031507.HR_964_spyact.shtml].

[18] H. Rep. 110-169. Available online at [http://www.congress.gov/cgi-lis/cpquery/R?cp110:FLD010:@1(hr169)].

[19] Information on this hearing, including a list of witnesses, witness testimony, and a link to the hearing webcast are available online at [http://judiciary.house.gov/Hearings.aspx?ID=170].

[20] A protected computer is defined in this bill as "a computer exclusively for the use of a financial institution or the U.S. government.

[21] For example, a Social Security number or other government-issued identification number, a bank or credit card number, or an associated password or access code.

[22] H. Rep. 110-169. Available online at [http://www.congress.gov/cgi-lis/cpquery/ R?cp110:FLD010:@1(hr169)].

[23] A protected computer is defined in this bill as " a computer used in interstate or foreign commerce or communication."

In: Computer Networks
Editor: Raymond S. Grigello

ISBN: 978-1-61209-596-7

Chapter 6

THE GOOGLE LIBRARY PROJECT: IS DIGITIZATION FOR PURPOSES OF ONLINE INDEXING FAIR USE UNDER COPYRIGHT LAW?*

Kate M. Manuel

ABSTRACT

The Google Book Search Library Project, announced in December 2004, raised important questions about infringing reproduction and fair use under copyright law. Google planned to digitize, index, and display "snippets" of print books in the collections of five major libraries without the permission of the books' copyright holders, if any. Authors and publishers owning copyrights to these books sued Google in September and October 2005, seeking to enjoin and recover damages for Google's alleged infringement of their exclusive rights to reproduce and publicly display their works. Google and proponents of its Library Project disputed these allegations. They essentially contended that Google's proposed uses were not infringing because Google allowed rights holders to "opt out" of having their books digitized or indexed. They also argued that, even if Google's proposed uses were infringing, they constituted fair uses under copyright law.

* This is an edited, excerpted and augmented edition of a CRS Report, Report R40194 dated February 5, 2009.

The arguments of the parties and their supporters highlighted several questions of first impression. First, does an entity conducting an unauthorized digitization and indexing project avoid committing copyright infringement by offering rights holders the opportunity to "opt out," or request removal or exclusion of their content? Is requiring rights holders to take steps to stop allegedly infringing digitization and indexing like requiring rights holders to use meta-tags to keep search engines from indexing online content? Or do rights holders employ sufficient measures to keep their books from being digitized and indexed online by publishing in print? Second, can unauthorized digitization, indexing, and display of "snippets" of print works constitute a fair use? Assuming unauthorized indexing and display of "snippets" are fair uses, can digitization claim to be a fair use on the grounds that apparently prima facie infringing activities that facilitate legitimate uses are fair uses?

On October 28, 2008, Google, authors, and publishers announced a proposed settlement, which, if approved by the court, could leave these and related questions unanswered. However, although a court granted preliminary approval to the settlement on November 17, 2008, final approval is still pending. Until final approval is granted, any rights holder belonging to the proposed settlement class—which includes "all persons having copyright interests in books" in the United States— could object to the agreement. The court could also reject the agreement as unfair, unreasonable, or inadequate. Moreover, even assuming final court approval, future cases may raise similar questions about infringing reproduction and fair use.

INTRODUCTION

Authors and publishers sued Google Inc. in 2005, shortly after Google announced plans to digitize books in the collections of several major libraries, index them in its search engine (http://www.google.com), and allow searchers to view "snippets" of the digitized books. Google's proposed reproduction and display of copyrighted books was not authorized by the rights holders, who alleged that the Google Library Project infringed their copyrights. Google's counterarguments—that allowing rights holders to "opt out" of having their books digitized or indexed kept its proposed uses from being infringing, or that, if found to be infringing, its proposed uses were fair—raised important questions about reproduction and fair use under copyright law. Namely, does an entity engaged in unauthorized digitization and indexing avoid committing copyright infringement by offering rights holders the opportunity to request removal or exclusion of their content? And, assuming unauthorized indexing and display of "snippets" are fair uses, can digitization claim to be a fair use on the grounds that apparently prima facie infringing activities that facilitate legitimate uses are fair uses? The proposed settlement agreement between Google and rights holders could mean that litigation over the Library Project does not help to answer these questions. However, final court approval of the settlement is still pending, and future digitization and indexing projects may raise similar questions.

This report provides background on the Library Project, legal issues raised by digitization and indexing projects, and the proposed settlement. It will be updated as developments warrant. It supersedes CRS Report RS22356, The Google Book Search Project: Is Online Indexing a Fair Use Under Copyright Law?, by Robin Jeweler.

THE GOOGLE LIBRARY PROJECT

In December 2004, Google initiated its Library Project by announcing partnerships with five libraries [1]. Under the partnership agreements, the libraries would allow Google to digitize the print books in their collections, and Google would (1) index the contents of the books; (2) display at least "snippets" of the books among its search results; and (3) provide partner libraries with digital copies of the print books in their collections [2]. Google and its partners never planned to make the full text of any digitized and indexed books that are still within their terms of copyright protection available to searchers [3]. Rather, by digitizing and indexing books, Google and its partners sought to make the contents of print books more accessible to searchers, who could potentially buy or borrow books after seeing "snippets" of them among the results of Google searches [4]. Google also intended to sell advertising "keyed" to results lists incorporating the digitized books [5].

Google's Library Project was itself part of a larger initiative initially known as Google Print and later renamed Google Book Search [6]. The Google Partner Program was also part of this initiative [7]. The Partner Program allowed authors and publishers to submit copies of their books for indexing in Google's search engine [8]. However, because rights holders affirmatively chose to have their books digitized or indexed through the Partner Program, the Program was not subject to allegations of copyright infringement like those made against the Library Project.

THE LITIGATION AND THE PARTIES' POSITIONS

Authors and publishers objected to the Google Library Project from its inception on the grounds that it infringed their copyrights.9 Generally, copyrights in books initially vest in the books' authors [10]. Many authors later transfer their copyrights to publishers under contract in exchange for payment and the publisher's manufacturing and selling copies of the book [11]. Regardless of whether they are the books' authors or publishers, however, copyright holders have exclusive rights "to reproduce the copyrighted work in copies," or, in the case of literary works such as books, "to display the copyrighted work publicly."[12]. The authors and publishers who objected to the Library Project claimed that Google infringed these exclusive rights by making digital copies of print books and presenting snippets from the digitized books without rights holders' permission [13]. Google initially responded to these concerns by allowing rights holders who did not want their books included in Google Book Search to "opt out."[14] If rights holders notified Google, Google would ensure that digitized versions of their books were not included in its database.

The ability to "opt out" of the Library Project did not satisfy authors and publishers, however. They sued to enjoin Google's digitization and indexing and to recover monetary damages for Google's alleged copyright infringement. In September 2005, the Authors Guild filed a class action suit in U.S. District Court for the Southern District of New York on behalf of "all persons or entities that hold the copyright to a literary work that is contained in the library of the University of Michigan."[15]. Shortly thereafter, five publishing companies also sued in the Southern District of New York [16]. The suits were consolidated, and additional plaintiffs, including the Association of American Publishers, joined the suit. Because the consolidated case was a class action, the court must approve any settlement of it [17].

In responding to the suit, Google essentially contended that its conduct was not infringing because it gave rights holders the opportunity to "opt out" of having their books digitized and indexed [18]. Google also claimed that, even if a court found its conduct to be infringing, this conduct represented a fair use of the rights holders' works [19]. Google and supporters of its Library Project specifically cited the decision by the U.S. Court of Appeals for the Ninth Circuit in Kelly v. Arriba Soft Corporation as support for the proposition that the indexing activities of Internet search engines constitute fair uses [20].

LEGAL ISSUES RAISED BY THE LITIGATION

The litigation over the Google Library Project raised important questions about infringing reproduction and fair use under copyright law. Namely, can an entity engaging in unauthorized digitization and indexing avoid liability for copyright infringement by offering rights holders the opportunity to request removal or exclusion of their content from its database? And, assuming unauthorized indexing and display of "snippets" of digitized works are fair uses, can digitization itself claim to be a fair use on the grounds that apparently *prima facie* infringing activities that facilitate legitimate uses are fair uses? These questions will arguably persist, and their answers remain important, even if the parties ultimately settle the litigation over the Library Project.

"Opt Out" Programs and Liability for Infringement

Google's first line of defense against the authors and publishers was essentially that it was not liable for copyright infringement because it gave rights holders the opportunity to "opt out" of having their works digitized and indexed. In making this argument, Google relied on the related claim that no one would conduct multi-library digitization and indexing projects like the Library Project if they had to clear the copyrights for every book with the rights holders [21]. Identifying and locating the rights holder(s) for one book can be difficult enough, supporters of the Google Library Project noted, without repeating this process millions of times, as would be necessary with a major library collection [22]. The publishers, in contrast, noted that Google's offer to let rights holders "opt out" of having their books digitized and indexed "stands copyright law on its head." [23]. They argued that one cannot generally announce one's intention to infringe multiple copyrighted works and collectively offer rights holders the opportunity not to have their works infringed [24].

It is impossible to predict what a court would find based on such arguments, and this report does not attempt to do so. This report does, however, highlight some of the considerations that could factor in the court's consideration of the issue. On the one hand, the requirement that a copyright owner act affirmatively to stop non-willful infringement is not without precedent. The "notice and takedown" procedures of the Digital Millennium Copyright Act (DMCA), [25] for example, require content owners to notify

Internet Service Providers (ISPs) of the existence of infringing content and can immunize ISPs from liability for infringement when they serve as "passive conduits" for infringing content transmitted by third parties [26] Similarly, at least one court has found that content owners are responsible for taking affirmative measures, such as using meta-tags within the computer code of a Web page, to prevent Internet search engines from automatically indexing and displaying their content [27]. On the other hand, plaintiffs could argue that comprehensive digitization projects, like that proposed by Google, willfully infringe copyright [28] and differ from the "passive conduits" protected by the DMCA. Likewise, rights holders in print books could argue that their situations differ from that of Web page authors because Google had to digitize their books before indexing them. They could claim that they took sufficient affirmative measures to protect their works by not making them available for free on the Web [29].

Digitization, Indexing, and Display as Fair Uses

Google also attempted to defend against the rights holders' allegations of copyright infringement by claiming that the Library Project, if found to be infringing, constituted a fair use [30]. The "fair use" exemption within copyright law limits rights holders' exclusive rights by providing that uses for "certain purposes"—including, but not limited to, criticism, comment, news reporting, teaching, scholarship, and research—do not infringe copyright even if they are made without the rights holders' consent [31]. In determining whether challenged conduct constitutes a fair use, a court considers the following factors, which were developed under the common law and later codified in the Copyright Act of 1976:

(1) the purpose and character of the use, including whether such use is of a commercial nature or is for nonprofit educational purposes;
(2) the nature of the copyrighted work;
(3) the amount and substantiality of the portion used in relation to the copyrighted work as a whole; and
(4) the effect of the use upon the potential market for or value of the copyrighted work [32]

These four factors must not be "treated in isolation, one from another." [33] Rather, "[a]ll are to be explored, and the results weighed together, in light of the purposes of copyright," [34] which is to "Promote the Progress of Science and useful Arts" and serve the public welfare.35 Also, because fair use is an "equitable rule of reason" to be applied in light of copyright law's overall purposes, other relevant factors may be considered.36 The court hearing the case makes findings of fact and assigns relative value and weight to each of the fair use factors. The court can also look to prior cases for guidance even though determining whether a challenged activity constitutes a fair use "calls for a case-by-case analysis."[37]

Although it is impossible to predict what a court would find when confronted with an actual case, and this report will not attempt to do so, it does highlight some of the many questions that the Google Library Project raised regarding each of the four statutory "fair use" factors. The report does so in order to illustrate the potential importance of the Library Project—or similar digitization and indexing projects—in establishing the scope of infringing reproduction and fair use under copyright law.

The Purpose and Character of the Use

First, as regards the purpose and character of the use, copyright law generally presumes that commercial uses are not fair, [38] and that transporting a work to a new medium is not a fair use [39] These presumptions would seem to work against digitization and indexing projects like the Library Project. The Project was implemented by a for-profit corporation that proposed, among other things, to sell ads "keyed" to the digitized content. The Project was also intended to migrate content from print to digital format. These presumptions can, however, be overridden when the use is sufficiently transformative [40]. A copy's use of the original is transformative when the copy does not "merely supersede[]" the original but rather "adds something new, with a further purpose or a different character" to the original [41].

The transformative nature of the Library Project would arguably be more easily established if it merely indexed books and displayed "snippets" of them. Were Google's uses so limited, it could probably rely on the precedent of two cases from the U.S. Court of Appeals for the Ninth Circuit which found that indexing and abridged displays of copyrighted content were fair uses. In the first case, *Kelly v. Arriba Soft Corporation*, the court held that a company operating a search engine, which had indexed a rights holder's online photographs and displayed "thumbnail" versions of them, was not liable for copyright infringement because its uses were fair [42]. Key to this holding was the court's finding that indexing represented a transformative use of the original photographs. While the original photographs were intended "to inform and to engage the viewer in an aesthetic experience," Arriba used its copies of them for a different function: "improving access to information on the internet" [43]. The court also emphasized that Arriba indexed and displayed "thumbnail" versions of the photographs [44]. The thumbnails had much lower resolution than the originals and thus could not substitute for them because "enlarging them sacrifices their clarity." [45] The Ninth Circuit reached a similar conclusion in Perfect 10, Inc. v. Amazon.com, Inc. [46]. There, the court also considered a use's benefit to society in finding the use to be transformative. The court noted that "a search engine provides social benefit by incorporating an original work into a new work, namely, an electronic reference tool." [47]

The digitization involved in the Library Project complicates the analysis, however. Admittedly, the prior cases that found indexing and abridged displays of copyrighted content to be fair uses also involved copying of originals.48 However, in these cases, the copying was of originals posted on the Internet and resulted in copies that were "inferior" to the originals for all purposes except their use in indexing. The first difference is potentially significant because courts have held that rights holders confer limited licenses to copy their content for purposes of indexing and abridged display by posting it on the Internet without taking affirmative measures to prevent copying [49]. The second difference could also be significant because digitized books are arguably superior to print ones when it comes to locating specific information within them [50].

Because digitization was so central to the Library Project, and arguably could not be directly paralleled to the copying in cases involving indexing and display of Internet materials, Google might have had to rely on the proposition that apparently prima facie infringing activities (such as digitization) that facilitate legitimate uses (such as indexing and limited displays) are fair uses. The Supreme Court's decision in Sony Corporation of America v. Universal City Studios could arguably provide broad support for this principle [51]. In Sony, the Court held that the sale of the video recording machine, which was used to "time shift" broadcast television for personal home viewing, was not contributory copyright infringement [52]. Although the factual underpinnings and legal precedent of Sony are not particularly relevant to or controlling in a case like Google's, the Sony decision itself stands as a landmark in copyright law demonstrating the willingness of the Court to balance new technological capabilities against traditional principles of copyright law and to recognize new categories of fair use. Many copyright experts saw analogies to the technological considerations inherent in Sony in Google's case [53]. Such experts noted that Google's allegedly infringing activity in digitizing print books was incidental to the valid and socially useful function of indexing.

The analogy to Sony might not be enough to persuade a court that digitizing for purposes of non- infringing indexing constitutes a fair use, however. Digitizing and indexing print books are arguably far removed from making and selling devices that consumers use to record broadcast television programming and replay it later. Additionally, courts have shown little inclination to recognize categories of judicially created fair uses other than time shifting. In UMG Recordings v. MP3.com, Inc., for example, a U.S. district court rejected out-of-hand the defendant's proffered fair use defense as a justification for unauthorized copying of plaintiffs' audio CDs [54]. The defendant had claimed that its unauthorized copying enabled CD owners to "*space shift*" because they could access the music on their CDs from any location through MP3.com's subscription service [55].

The Nature of the Copyrighted Work

Comprehensive digitization and indexing projects, such as the Google Library Project, raise similar questions when the second fair use factor is considered. Projects that digitize library collections potentially encompass diverse types of materials. Some of these materials may be works of fiction, which are among the creative works accorded the highest level of copyright protection [56]. Other materials may be reference books or compendiums of facts, which are afforded the "thinnest" copyright protection [57]. Yet other materials may be nonfiction and mix unprotected ideas with protected expressions of these ideas [58]. This diversity of materials makes possible the arguments of both proponents and opponents of the view that projects like Google Book Search constitute fair uses. The nature of the work can, however, be less important than the purpose and character of the use, at least in situations where the use can be clearly recognized as transformative [59].

The Amount and Substantiality of the Portion Used

The amount and substantiality of the portion used in relation to the copyrighted work as a whole is another factor that could potentially cut either way in cases involving digitization and indexing projects. As a general rule, "[w]hile wholesale copying does not preclude fair use per se, copying an entire work militates against a finding of fair use." [60] Copying entire works can, however, be found to constitute a fair use when doing so is reasonable given the purpose and character of the use [61]. Digitization projects, such as the Google Library Project, would clearly be engaged in wholesale copying, including copying any segments comprising the "heart" of the copied work [62]. The question would thus become whether such wholesale copying was reasonable for an indexing project. Proponents of the project could argue that courts have found copying entire works in order to digitize them reasonable, [63] and that searchers would see only "snippets" of the work in any case. Opponents, in contrast, could argue that, in all cases where courts protected wholesale copying for purposes of indexing, the authors had placed their works online, thereby creating implied licenses for others to copy and index them [64].Moreover, in at least some of these cases, the copies were deleted after the indexing was completed [65]. In no case did the copier propose to give copies to third parties, as Google did when contracting to provide digital copies of the books in their collections to libraries.

The Effect of the Use Upon the Potential Market or Value of the Work

Finally, digitization and indexing projects could be seen as either promoting or inhibiting the potential markets or values of the copyrighted works. Proponents of digitization could argue that indexing and display of "snippets" of print books increases the markets for the originals by alerting researchers to books on their topics.

If researchers purchase books of which they would otherwise have been unaware, the markets for these books could potentially be improved by the unauthorized digitization. Opponents, in contrast, could argue that unauthorized digitization and indexing usurps markets that the rights holders are developing; [66] that viewing "snippets" of print books sometimes can substitute for purchases of them; and that rights holders should be free to determine whether, when, and how their print works are digitized [67]. The outcome of any findings by the court on this factor may hinge upon the degree of harm to their markets that plaintiffs must show.

Some courts have required plaintiffs to show only that the markets in which they alleged harm are "likely to be developed," [68] while others have required proof of actual losses in established markets [69]. The fact that a use is transformative can, however, outweigh even inhibition of or harm to plaintiffs' markets [70].

The Proposed Settlement Agreement

On October 28, 2008, Google and the rights holders announced a proposed settlement agreement [71]. Under this agreement, Google would compensate rights holders for prior and future uses of their work [72]. Google would also fund the establishment and initial operations of a not-forprofit entity, called the Registry, which would represent rights holders in negotiating future uses of their content with Google [73]. Google, in turn, would receive a non-exclusive license74 to (1) "Digitize all Books and Inserts" published before January 5, 2009, and (2) make certain uses of the digitized materials, including displaying "snippets" of them among its search results, subject to the terms of the agreement [75]. By allowing Google to digitize and display books, the agreement would pave the way for Google to expand Google Book Search, selling subscriptions to institutions and electronic versions of books to individuals [76]. The agreement would also create certain rights and responsibilities for libraries that allow Google to digitize their books, [77] as well as make certain provisions for institutional subscribers to, or individual users of, commercialized versions of Google's Book Search database [78]. The agreement will not take effect until certain conditions are met, one of which requires final court approval of the settlement agreement [79]. The court granted preliminary approval of the agreement on November 17, 2008 [80]. Final approval is, however, still pending. Class members presently have until May 5, 2009, to file objections with the court [81]. At least some objections will probably be filed because the proposed settlement class is both broad and diverse. It encompasses "all persons having copyright interests in books" under U.S. law [82] and includes authors working in different genres (fiction, non-fiction, textbooks, anthologies, reference works, etc.), some of whom have expressed dissatisfaction with the agreement [83]. Moreover, because the suit is a class action, the judge is required, under the Federal Rules of Civil Procedure, to review the proposed settlement to ensure that it is "fair, reasonable, and adequate." [84]. The judge could reject the proposed settlement based upon concerns expressed by its critics, who fear monopolization of the market by Google or the Registry, among other things, [85] or based upon other concerns [86]. Rejection of the proposed settlement agreement could place the parties' claims and defenses back before the court [87].

Acknowledgments

Robin Jeweler, Legislative Attorney, original author of RS22356 The Google Book Search Project: Is Online Indexing a Fair Use Under Copyright Law?

References

[1] Google Checks Out Library Books, Dec. 14, 2004, available at http://www.google.com/press/pressrel/ print_library.html. Participating libraries included those at the University of Michigan, Harvard University, Stanford University, and Oxford University, as well as the New York Public Library.
[2] *Id.*
[3] *Id.* Copyright protection for books generally lasts "for a term consisting of the life of the author and 70 years after the author's death." 17 U.S.C. § 302(a).
[4] Google Checks Out Library Books, supra note 1.
[5] *Id.*
[6] Association of American University Presses, Google Book Search, Neé Google Print, available at http://www.aaupnet.org/aboutup /issues /gprint.html.
[7] Google Books Partner Program: Promote Your Books on Google—For Free, 2009, available at http://books.google.com/googlebooks/ boo k_search_tour.
[8] *Id.*
[9] *See, e.g.*, Anandashankar Mazumdar, University Press Group Expresses Concern Over Google Print's Digitization of Works, 70 *Pat., Trademark and Copyright J.* 109 (June 3, 2005).
[10] 17 U.S.C. § 201(a). There are exceptions to this general rule, such as when a book is "made for hire" or is a "work of the United States Government." See 17 U.S.C. § 105 and 201(b).
[11] See, e.g., Example Author Contract, available at http://www.writ econtent.com/Publishing_Tools/Author_Contract_/author_contract_.htm l ("The Author hereby grants to the Publisher exclusive rights to reproduce and/or publish or adapt and sell, and/or license third parties to publish or adapt and sell said Work.").
[12] 17 U.S.C. § 106(1) and (5).

[13] *See, e.g.*, Mazumdar, *supra* note 9.

[14] *See, e.g.*, Christine Mumford, Google Library Project Temporarily Halted to Allow Copyright Owner Response, 70 *Pat., Trademark and Copyright J.* 461 (Aug. 19, 2005).

[15] Authors Guild v. Google Inc., Class Action Complaint, No. 05 CV 8136 (S.D.N.Y. Sept. 20, 2005) at 20. The University of Michigan's library was the focus because Google began digitizing its books first. Id. at 31. Under copyright law, "literary works" are any "works, other than audiovisual works, expressed in words." 17 U.S.C. § 101.

[16] McGraw Hill Cos. v. Google Inc., Complaint, No. 05 CV 8881 (S.D.N.Y. Oct. 19, 2005). These companies were McGraw-Hill Companies; Pearson Education; Penguin Group; Simon and Schuster; and John Wiley and Sons.

[17] Fed. R. Civ. P. 23(e).

[18] *See,* e.g., Susan Wojcicki, Google Print and the Authors Guild, Sept. 20, 2005, available at http://googleblog.blogspot.com/2005/09/google-print-and-authors-guild.html.

[19] *Id.*

[20] *See* 336 F.3d 811 (9th Cir. 2003). For more background on Kelly, see CRS Report RL33810, *Internet Search Engines: Copyright's "Fair Use" in Reproduction and Public Display Rights*, by Robin Jeweler and Brian T. Yeh.

[21] *See, e.g.*, Wojcicki, *supra* note 18.

[22] The Harvard University Libraries (HUL), for example, contain over 15 million books. HUL, About the HOLLIS Catalog, June 25, 2007, available at http://lib.harvard.edu/catalogs/hollis.html.

[23] Anandashankar Mazumdar, Publishers: Value of Book Search Project Shows That Scanning Is Not Fair Use, 71 *Pat., Trademark and Copyright J.* 94 (Nov. 25. 2005).

[24] *Id.*

[25] P.L. 105-304. For more information on the DMCA generally, see CRS Report 98-943, *Digital Millennium Copyright Act, P.L. 105-304: Summary and Analysis*, by Dorothy M. Schrader.

[26] 11 U.S.C. § 512(b)-(c).

[27] *See* Field v. Google Inc., 412 F. Supp. 2d 1106 (D. Nev. 2006).

[28] *See, e.g.,* Class Action Complaint, supra note 15, at 23.d and 41 (alleging Google's infringement was willful); Complaint, supra note 16, at 2 (same).

[29] See, e.g., Complaint, supra note 16, at 29 (arguing that Web pages differ from print books because rights holders in Web pages can rely on technological measures to prevent indexing, while authors of print books can take no such measures to prevent digitization).
[30] *See, e.g.*, Wojcicki, supra note 18.
[31] 17 U.S.C. § 107.
[32] *Id.*
[33] Campbell v. Acuff-Rose Music, Inc., 510 U.S. 569, 578 (1994).
[34] *Id.*
[35] Perfect 10, Inc. v. Amazon.com, Inc., 487 F.3d 701, 720 (9th Cir. 2007) (quoting the U.S. Constitution, art. I, § 8, cl. 8, as well as *Sony Corp. of Am. v. Universal City Studios, Inc.*, 464 U.S. 417, 429 n.10 (1984)).
[36] Stewart v. Abend, 495 U.S. 207, 237 (1990).
[37] Campbell, 510 U.S. at 577-78.
[38] See, e.g., Sony, 464 U.S. at 451 ("Every commercial use of copyrighted materials is presumptively an unfair exploitation of the monopoly privilege that belongs to the owner of the copyright.").
[39] See, e.g., Infinity Broad. Corp. v. Kirkwood, 150 F.3d 104, 108 (2d Cir. 1998) (retransmission of radio broadcast over telephone lines not a fair use); UMG Recordings, Inc. v. MP3.com, Inc., 92 F. Supp. 2d 349, 351 (S.D.N.Y. 2000) (reproducing analog audio CDs as MP3s not a fair use).
[40] Campbell, 510 U.S. at 578-79.
[41] *Id.* at 579.
[42] 336 F.3d 811 (9th Cir. 2003).
[43] *Id.* at 818-19.
[44] *Id.* at 818.
[45] *Id.* at 819.
[46] 487 F.3d 701, 721 (9th Cir. 2007), *rev'g* Perfect 10, Inc. v. Google Inc., 416 F. Supp. 2d 828 (C.D. Cal. 2006) (holding that Google's use of thumbnail versions of Perfect 10's copyrighted photographs was not fair, in part, because Google's thumbnails could potentially substitute for the reduced-size versions of these photographs that Perfect 10 had licensed another company to reproduce and distribute for display on cell phones).
[47] Perfect 10, Inc., 487 F.3d at 721.
[48] *See, e.g.*, Kelly, 336 F.3d at 816.
[49] Field, 412 F. Supp. 2d at 1115-16.
[50] A digital version of a print book would display poorer resolution than the original. However, it would enable researchers to locate specific

content more easily by using the "search" or "find" functions of their Web browsers.

[51] 464 U.S. 417 (1984).

[52] *Id.* at 442.

[53] *See,* e.g., Jonathan Band, The Google Print Library Project: Fair or Foul?, 9 J. of Internet L. 1, 4 (Oct. 2005); Christopher Heun, Courts Unlikely to Stop Google Book Copying, InternetWeek (Sept. 2, 2005), available at http://internetweek.cmp.com/showArticle.jhtml? articleID= 170700329.

[54] 92 F. Supp. 2d 349, 352 (S.D.N.Y. 2000) ("[D]efendant's 'fair use' defense is indefensible and must be denied as a matter of law.").

[55] *Id.*

[56] Kelly, 336 F.3d at 820.

[57] See, e.g., Feist Publ'ns, Inc. v. Rural Tel. Serv., 499 U.S. 340 (1991) (requiring originality in the selection or arrangement of facts for copyrightability).

[58] See, e.g., Baker v. Selden, 101 U.S. 99 (1879) (distinguishing non-protectable ideas from their protectable expressions).

[59] Campbell, 510 U.S. at 577-78.

[60] Worldwide Church of God v. Philadelphia Church of God, Inc., 227 F.3d 1110, 1118 (9th Cir. 2000) (internal quotations omitted).

[61] *See, e.g.*, Kelly, 336 F.3d at 821; Sega Enters., Ltd. v. Accolade, Inc., 977 F.2d 1510, 1523 (9th Cir. 1992).

[62] *See, e.g.*, Harper and Row Publishers, Inc. v. Nation Enters., 471 U.S. 539, 564 (1985).

[63] *See, e.g.*, Kelly, 336 F.3d 811; Perfect 10, Inc., 487 F.3d 701.

[64] *See* Field, 412 F. Supp. 2d at 1115-16.

[65] *See* Kelly, 336 F.3d at 816.

[66] *See* Complaint, *supra* note 16, at 5 (noting that publishers were already making their print books available online in various ways, including a partnership with the search engine Yahoo!).

[67] *Cf.* BMG Music v. Gonzalez, 430 F.3d 888, 891 (7th Cir. 2005) ("Copyright law lets authors make their own decisions about how best to promote their works.").

[68] *See,* e.g., Bill Graham Archives, LLC v. Dorling Kindersley, Ltd., 386 F. Supp. 2d 324, 332 (S.D.N.Y. 2005).

[69] Perfect 10, Inc., 487 F.3d at 725.

[70] *See, e.g.*, Campbell, 510 U.S. at 591.

[71] Authors Guild, Inc. v. Google Inc., Settlement Agreement, Case No. 05 CV 8136-JES (S.D.N.Y. Oct. 28, 2008).

[72] Id. at 2.1(a) (providing that Google would pay 70% of the net revenue earned from uses of Google Book Search in the United States to rights holders); 2.1(b) (providing that Google would pay at least $45 million into a "Settlement Fund," whose proceeds would pay rights holders whose books or "inserts" were digitized prior to January 5, 2009).

[73] Id. at 2.1(c). Among other functions, the Registry could negotiate the terms of "New Revenue Models" (e.g., print on-demand) with Google and negotiate pricing categories and percentages for sale of digitized materials to users.

[74] Because this license is non-exclusive, the Registry could license other entities to digitize, index, or display the works of rights holders. However, if the Registry were to enter into a similar agreement within 10 years of the settlement's effective date, it must extend comparable economic and other terms to Google. *Id.* at 3.8(a).

[75] *Id.* at 3.1. Google's rights to use books within their terms of copyright protection would hinge upon whether they were "commercially available," or available "for sale new through one or more then-customary channels of trade in the United States." See id. at 1.28. If a book is commercially available, Google could not make "display uses" without the copyright holders' consent. Id. at 3.3-3.5. Conversely, if a book is not commercially available, Google could make "display uses" unless the rights holder objects. Id. This distinction between commercially available and non- commercially available books would significantly vary the legal protections of copyright law, which protects all works equally, regardless of their commercial availability, during their terms of copyright protection. *See* 17 U.S.C. § 106 and § 302.

[76] *See, e.g.*, Settlement Agreement, *supra* note 71, at 3.7.

[77] *Id.* at 7.2(f)(i)-(ii) and Article X.

[78] *See, e.g.*, James Grimmelmann, Principles and Recommendations for the Google Book Search Settlement, Nov. 8, 2008, available at http://www.laboratorium.net/archive/2008/11/08/principles_and_recom mendations_for_the_ google_book (noting the existence, as well as the potential inadequacy, of provisions regarding users).

[79] Settlement Agreement, *supra* note 71, at 1.49.

[80] Authors Guild, Inc. v. Google Inc., Order Granting Preliminary Settlement Approval, Case No. 05 CV 8136-JES (S.D.N.Y. Nov. 17,

2008). The final hearing is presently scheduled to be held on June 11, 2009.

[81] *Id.*

[82] Settlement Agreement, *supra* note 71, at 1.142.

[83] *See, e.g.*, Science Fiction and Fantasy Writers of America, Inc. (SFWA), SFWA Statement on Google/Author's Guild Settlement, Oct. 31, 2008, available at http://www.sfwa.org/news/2008/sfwastatement.htm.

[84] Fed. R. Civ. P. 23(e).

[85] *See, e.g.*, Grimmelman, *supra* note 78; James Gibson, Google's New Monopoly? How the Company Could Gain by Paying Millions in Copyright Fees, Wash. Post, Nov. 3, 2008, at A21.

[86] *See, e.g.,* Muchnick v. Thompson Corp., 509 F.3d 116 (2d Cir. 2007) (quashing the proposed settlement agreement resolving the litigation in Tasini v. New York Times because some members of the proposed settlement class had not registered their works with the U.S. Copyright Office and so lacked standing to bring suit in federal court). The district court had previously approved this agreement.

[87] Even if eventually approved by the courts, the settlement agreement only governs claims against Google over its Library Project within the United States. Litigation in other jurisdictions remains possible. *See, e.g.*, Editions du Seuil v. Google Inc., Tribunal de Grand Instance de Paris.

INDEX

A

access, vii, viii, ix, 1, 2, 3, 4, 5, 6, 8, 11, 12, 13, 15, 16, 18, 19, 20, 21, 24, 25, 27, 28, 29, 31, 32, 33, 35, 37, 38, 40, 41, 42, 43, 44, 46, 47, 54, 55, 56, 57, 58, 59, 60, 62, 63, 74, 75, 91, 106, 110, 118, 119
accessibility, 19
accuracy, 17
administration, 72, 73
administrative, 25, 28, 80, 85
administrators, 32
advertisements, xi, 99, 100, 102, 107
advertising, 101, 106, 107, 113
adware, xi, 99, 102, 103, 107, 109
aesthetic, 118
African American, 14
African Americans, 14
agencies, ix, x, 25, 28, 53, 54, 56, 59, 60, 65, 71, 74, 76, 80, 102
aggregate demand, 64
Air Force, 92
al Qaeda, 84
Alaska, 26, 28
alternative, 81
Amazon, 118, 124
amendments, 39
America COMPETES Act, 35
America Online, 103
American Recovery and Reinvestment Act, v, vii, viii, ix, 10, 12, 13, 17, 20, 37, 43, 45, 50, 53, 54, 55, 68
American Recovery and Reinvestment Act of 2009, 13, 20, 37, 55, 68
analog, 124
analysts, 76, 79
anticompetitive, viii, 2, 7
anti-spyware, 103, 104, 105, 106, 108
antitrust, 6, 7
appropriations, xi, 72, 87
Appropriations Act, 35, 36, 45
Appropriations Committee, 39, 66
ARC, 28
argument, 83, 95, 101, 115
armed forces, 85, 95, 96
assessment, 19, 45, 90
attachment, xi, 100, 102
attacks, x, 72, 73, 74, 75, 78, 80, 81, 82, 84, 85, 90, 91, 97
audits, 40, 43, 56, 66
Aurora, 91
authorities, x, 72, 79, 86, 89, 95
authority, 3, 5, 6, 8, 9, 28, 34, 59, 64, 66, 76, 77, 79, 80, 81, 82, 83, 85, 86, 89, 93, 94, 95, 96, 97, 98, 109
availability, ix, 12, 14, 17, 18, 34, 38, 51, 54, 55, 64, 65, 126
average costs, 30
averaging, 50
awareness, 56, 74, 79, 88

B

back, 102, 121
background information, 94
Barack Obama, 50
barriers, 18, 19
base, 30, 102
basic research, 35
behavior, viii, 2, 5, 6, 7
BellSouth, 4, 109
benchmarks, 60
beneficiaries, 31
benefits, ix, 5, 7, 23, 28, 54, 102
broadband, viii, ix, x, 2, 3, 4, 5, 6, 7, 8, 9, 10, 11, 12, 13, 14, 15, 16, 17, 18, 19, 20, 21, 22, 23, 27, 28, 30, 33, 34, 35, 36, 37, 38, 39, 40, 41, 42, 43, 44, 45, 46, 47, 48, 49, 51, 53, 54, 55, 56, 57, 58, 59, 60, 61, 62, 63, 64, 65, 66, 67, 68, 103
Broadband, viii, ix, 2, 8, 10, 11, 12, 13, 14, 16, 17, 18, 19, 20, 23, 27, 28, 30, 34, 36, 37, 38, 39, 40, 42, 43, 44, 45, 47, 48, 49, 50, 51, 53, 54, 55, 56, 58, 59, 63, 65, 66, 68
Broadband Data Improvement Act, ix, 17, 34, 36, 38, 40, 42, 43, 53, 56, 65
Broadband grants and loans, ix, 53
Broadband Technology Opportunities Program (BTOP), viii, ix, 2, 8, 53, 55, 56, 59
broadcast television, 118, 119
Bush Administration, x, 18, 19, 71, 73, 95
businesses, 6, 13, 15, 45, 54

C

cable modem, 3, 13, 15, 54, 63
cable system, 6, 63
cable television, 3, 7, 13
campaigns, 85, 86
Canada, 15
CAN-SPAM Act, 101
carrier, 3
Catalog of Federal Domestic Assistance, 26
category d, 12
CDBG, 27
CDT, 100, 101, 108
cell, 124
cell phones, 124
Census, 17, 34, 36, 65
Census Bureau, 17, 34, 36, 65
CEO, 9, 10, 49, 109
challenges, ix, x, 54, 55, 65, 71
City, 109, 118, 124
civil service, 80, 94
civilian, 76
clarity, 118
classification, 3
classrooms, 31
climate, 19
CNN, 91
Co, 15, 82, 95, 96
commerce, 9, 15, 49, 80, 110
commercial, 20, 54, 65, 101, 116, 117, 124, 126
Committee on Homeland Security, 87
Committee on Intelligence, 75, 90, 91, 92
Committee on the Judiciary, 106
common law, 116
communication, 110
Communications Act, 3, 8, 50
Communications Act of 1934, 3, 50
communities, 12, 14, 15, 17, 20, 23, 25, 26, 27, 33, 34, 36, 40, 65, 74, 78, 98
community, 19, 23, 27, 28, 35, 40, 41, 42, 43, 56, 64, 88, 89
Community Connect Grant Program, 45
Community Development Block Grant, 27
competition, ix, 3, 5, 6, 7, 12, 18, 19, 46, 54, 61, 62, 67
competitive advantage, 75
competitors, 5, 61
components, 74, 81, 91, 97, 103
Comprehensive National Cybersecurity Initiative, v, vii, x, 71, 73
Comprehensive National Cybersecurity Initiative (the CNCI), x, 71, 73
computer, vii, x, xi, 40, 42, 43, 56, 60, 71, 73, 75, 76, 80, 84, 91, 92, 99, 100, 101,

102, 103, 104, 105, 106, 107, 109, 110, 116
computer fraud, 80
computer software, 105
computer systems, 75, 80, 91
computer use, xi, 99, 103, 110
Conference Report, 55, 59, 63, 64, 68
confidence, 78
conflict, 2, 81, 83, 96
congress, 106, 110
connectivity, 20, 23, 27, 75
consensus, viii, 2, 5, 7, 108
consent, 6, 9, 100, 101, 103, 107, 116, 126
Consolidated Appropriations Act, 35
Consolidated Security, Disaster Assistance, and Continuing Appropriations Act of 2009, 98
consolidation, 2
Constitution, 79, 81, 82, 86, 89, 93, 94, 97, 124
constitutional law, 79
construction, 22, 55, 63
consumer choice, 5
consumer education, 101
consumer protection, 8
consumers, vii, 1, 2, 3, 4, 5, 15, 18, 33, 44, 56, 57, 61, 62, 63, 101, 102, 104, 105, 119
control, vii, 1, 2, 4, 5, 74, 75, 78, 82, 87, 88, 91, 100, 101, 105, 107
controversial, 61, 62, 63, 66
coordination, x, 40, 42, 44, 54, 57, 60, 62, 65, 72, 77, 78, 98
copper, 13, 54
copyright holders, xi, 111, 114, 126
copyright law, xi, 111, 112, 115, 116, 117, 118, 123, 126
copyrights, xii, 111, 112, 114, 115
corporations, 103
cost, 15, 22, 23, 30, 31, 33, 47, 50, 67
court approval, xii, 112, 113, 121
Court of Appeals, 4, 114, 117
courts, 82, 83, 94, 118, 119, 120, 127
covering, 32
covert action, 87, 88, 89, 97
credit, xi, 21, 41, 45, 46, 99, 102, 110
credit card, xi, 99, 102, 110
crime, 76
criminal activity, 84
critical infrastructure, 26, 75, 78, 91
criticism, 116
cross-country, 15
curriculum, 23
customers, 15, 27, 28
cyber attackers, x, 71
cyber security, 73, 77, 78, 79, 89, 93
cyber threats, x, 71, 72, 73, 76, 79, 80, 83, 87, 89
cyber-attack, 90
cybercrime, 79, 94
cyber-crime, 76
cyber-intrusions, x, 71
cybersecurity, x, 71, 72, 73, 76, 77, 79, 80, 82, 83, 84, 85, 86, 87, 88, 89
cybersecurity arena, xi, 72, 84
cyberspace, 73, 76, 79

D

data collection, x, 16, 17, 36, 47, 54
data gathering, 16
database, 67, 69, 75, 114, 115, 121
death, 122
debates, 33
decisions, x, 39, 54, 59, 60, 64, 67, 125
defense, 88, 92, 115, 119, 125
defenses, 76, 121
definition, vii, 1, 2, 18, 31, 33, 34, 51, 61, 87, 95, 100, 103
defraud, 106
degradation, 91
delivery, 4, 7, 24, 29, 106
Department of Agriculture, ix, 20, 27, 28, 35, 37, 53, 55, 68
Department of Commerce, ix, 15, 17, 27, 28, 29, 35, 36, 37, 40, 43, 48, 50, 53, 55, 65, 66, 68
Department of Defense, 90, 92
Department of Education, 29
Department of Energy, 74

Department of Health and Human Services, 24, 29
Department of Homeland Security, 79, 90, 91, 93, 109
Department of Housing and Urban Development, 27
Department of Justice, 100
deregulation, 18
destruction, 74, 75
detainees, 96
detection, 77
detection system, 77
detention, 84, 97
DHS, 79, 88, 90, 93
dial-up, viii, 11, 103
digital divide, vii, viii, ix, 11, 12, 13, 18, 19, 21, 46
digitization, xii, 112, 113, 114, 115, 116, 117, 118, 119, 120, 124
directives, 73, 77, 93
Director of National Intelligence, 75, 91
discrimination, 7, 8, 44, 57, 62
discriminatory, vii, 1, 2, 6, 7
disputes, 83, 94, 95
dissatisfaction, 121
distance learning, 21, 28, 63
distribution, 102
District of Columbia, 32, 44, 57, 61, 64
diversification, 2
diversity, 16, 88, 119
division, 44, 57, 64, 96
DOC, viii, ix, 8, 9, 12, 41, 43, 46, 49, 53, 55, 56, 58
domestic issues, 46
download, 13, 17, 41, 42, 61, 63, 104
drawing, 82
DSL, 3, 13, 15, 27, 54
due process, 96
duties, 76, 88

E

e-commerce, 15
economic development, 15, 21, 25, 26, 37, 40, 43, 58, 60
Economic Development Administration, 15
economic growth, 15, 18, 27, 56
economic security, 74
economic stimulus package, viii, 2, 7, 37, 55
economically disadvantaged, 45, 57
education, 12, 21, 24, 31, 32, 37, 55, 56, 77, 81, 101
Education, 23, 24, 28, 29, 123
electrical power, 75
electricity, 74
electronic surveillance, 82, 84
emergency, 33, 39, 82, 95
emergency response, 39
employees, 91
employment, 15
empowered, 80
encryption, 109
enemies, 72
enemy combatan, 97
enemy combatants, 97
energy, 74, 75
energy pipelines, 75
enforcement, 3, 7, 62, 78, 80, 87, 98, 100, 101, 108
enterprise, 77
environment, 19, 72
equipment, 23, 27, 37, 41, 54, 55, 56, 74
espionage, 80
Estonia, 73
evidence, 109
exclusion, xii, 112, 115
executive branch, 73, 81, 82, 85, 88, 89
Executive Branch, 79
executive orders, 77
executive power, x, 72, 80, 94
exercise, 82, 84, 92, 93
expenditures, 41
expertise, 59, 78
exploitation, 75, 124
external relations, 96

F

failure, 5
fair use, xi, xii, 111, 112, 114, 115, 116, 117, 118, 119, 124, 125

FCC, vii, viii, ix, 2, 3, 4, 5, 6, 8, 9, 10, 13, 14, 16, 17, 18, 20, 30, 31, 32, 33, 34, 35, 36, 38, 39, 40, 42, 43, 44, 45, 46, 47, 49, 50, 51, 53, 55, 56, 57, 59, 61, 62, 65, 67, 68
February, viii, x, 4, 9, 10, 12, 15, 28, 39, 41, 43, 45, 48, 55, 68, 72, 73, 93, 111
federal agency, 76, 92
federal assistance, ix, 12, 46, 54, 67
Federal Communications Commission, vii, ix, 2, 3, 9, 10, 27, 28, 30, 47, 49, 53, 55
federal criminal law, 80
federal government, viii, x, 12, 13, 18, 19, 21, 71, 73, 77, 78, 93
federal law, xi, 100
Federal Register, 8, 68
Federal Trade Commission, xi, 100, 102
Federal Trade Commission (FTC), xi, 100, 102
fees, 5, 8
fiber, viii, 11, 13, 27, 28, 54, 63
finance, 6, 37, 41
financial, viii, 12, 13, 19, 20, 21, 34, 40, 42, 45, 54, 58, 60, 75, 78, 87, 102, 110
financial institution, 110
financing, 22, 27, 46
first generation, 47
foreign affairs, x, 72, 81, 82, 85, 88, 97
Foreign Intelligence Surveillance Act, 82, 84, 94
foreign nation, x, 71, 96
fraud, 66, 80
FTC, 100, 101, 104, 105, 106, 109
fuel, 101
funding, ix, 20, 27, 28, 31, 32, 40, 41, 42, 43, 50, 51, 53, 56, 58, 61, 63, 65, 68, 77, 89, 90
funds, 24, 29, 30, 32, 34, 36, 37, 40, 42, 43, 44, 57, 58, 59, 60, 63, 66, 67, 87

G

generation, 13, 19, 20, 41, 44, 46, 57, 63
goals, ix, x, 34, 46, 50, 51, 54, 60, 63, 65, 67, 78, 87
google, 112, 122, 123, 126
Google Book Search Library Project, xi, 111
Google Library Project, v, vii, 111, 112, 113, 114, 115, 117, 119, 123
government, viii, ix, x, xi, 12, 13, 18, 19, 20, 21, 37, 42, 46, 48, 54, 55, 61, 65, 67, 71, 72, 73, 74, 76, 77, 78, 80, 81, 83, 88, 91, 93, 97, 102, 109, 110
Government Accountability Office, 16, 17, 34, 36, 49, 65
Government Accountability Office (GAO), 16
government computer networks, x, 71
Government Computer News, 93
government intervention, ix, 12, 19, 46, 54, 67
governments, 42
GPO, 68
grant programs, ix, 39, 40, 41, 53, 54
grants, ix, 13, 17, 19, 22, 23, 25, 26, 27, 28, 35, 36, 37, 38, 39, 40, 41, 42, 43, 44, 45, 46, 53, 55, 56, 57, 58, 59, 60, 61, 62, 63, 64, 65, 66, 67, 122
greed, 39
grids, 74, 75
groups, x, 37, 55, 71, 76, 102, 108
growth, 7, 14, 15, 18, 27, 56, 100
guidance, 97, 117
guidelines, vii, 1, 5, 44, 56, 59, 73

H

hackers, 73, 84
harm, 3, 5, 62, 74, 79, 120
Harvard, 122, 123
Hawaii, 105
health, 21, 23, 24, 28, 29, 31, 32, 74
Health and Human Services, 24, 28, 29
health care, 21, 23, 28, 31, 32
hearing, 7, 9, 10, 31, 87, 105, 106, 108, 110, 117, 127
heart, 120
high income areas, viii, 11, 12, 14
high-speed, viii, 2, 10, 11, 14, 16, 18, 21, 28, 46, 54, 58
homeland security, 74, 78

Homeland Security, 73, 88, 90, 91, 92, 93, 98, 106
House, xi, 7, 35, 36, 37, 39, 41, 42, 43, 45, 50, 55, 66, 76, 87, 90, 91, 93, 100, 105, 106, 107, 108, 110
House Appropriations Committee, 37
House of Representatives, xi, 100, 105
household, 101
households, 13
Housing and Urban Development, 27

I

ICT, 48
identification, 110
identity, 102, 109
implementation, ix, x, 44, 45, 54, 55, 56, 60, 61, 62, 65, 67, 77
improvements, 27
in situ, 119
incentive, 2, 9, 15
incentives, 19, 20, 46
income, viii, 11, 12, 14, 16, 20, 27, 30, 33, 45, 46, 55
income tax, 45
income-based programs, 30
incomes, 14, 15
increased access, 19
incumbents, 64
indexing, xii, 112, 113, 114, 115, 116, 117, 118, 119, 120, 124
indexing project, xii, 112, 113, 115, 117, 119, 120
Indian, 27, 41, 42, 44, 57, 64
individual keystrokes, xi, 99
individuals, x, xi, 12, 15, 30, 31, 71, 72, 73, 75, 121
industries, 98
industry, 4, 16, 54, 83, 89, 91, 100, 101, 105
inertia, 83, 85
Infiltration, 109
information exchange, 78, 88
information systems, 72, 74, 75
information technology, 23, 28, 46, 54, 77, 103
Information Technology, 92
infrastructure, vii, 10, 18, 20, 22, 25, 26, 27, 37, 41, 44, 46, 54, 55, 57, 64, 67, 73, 74, 75, 78, 84, 91, 92
infringement, xii, 111, 112, 113, 114, 115, 116, 117, 118, 123
inhibition, 120
innovation, 2, 6, 19, 20, 50
Innovation, 27, 49, 50
insight, 73, 77
Inspector General, 40, 43, 56, 66
institutions, 13, 20, 41, 54, 56, 121
instruments, 80
intelligence, x, 71, 72, 76, 77, 82, 84, 85, 86, 87, 88, 89, 94, 96, 98
Intelligence Authorization Act, 97
Intelligence Community, 75, 91, 92
intelligence gathering, 84, 88
Internal Revenue Code, 45
international law, xi, 72, 94
international relations, 86, 97
internet, 74, 78, 79, 80, 92, 118
Internet, vii, viii, xi, 1, 2, 3, 4, 5, 6, 7, 8, 9, 10, 11, 12, 13, 14, 15, 16, 19, 27, 31, 32, 33, 34, 44, 46, 47, 48, 49, 51, 54, 57, 62, 67, 68, 75, 92, 99, 101, 102, 104, 106, 108, 109, 110, 114, 116, 118, 123, 125
Internet Protocol, 49
interstate, 3, 80, 110
interstate commerce, 80
intervention, ix, 12, 18, 19, 46, 54, 67
intrinsic, 77
intrusions, x, 71, 76, 78, 80, 82, 84, 85
investment, ix, 7, 12, 18, 19, 20, 27, 32, 41, 42, 45, 46, 54, 67

J

job creation, ix, 44, 54, 56, 57, 63, 67
jobs, 15, 46, 55
judge, 109, 121
judges, 59
judgment, 86
judiciary, 110
Judiciary, 7, 106
jurisdiction, 3, 8, 88
jurisdictions, 127

justification, 88, 119

K

Kentucky, 17, 28
keylogging, xi, 99, 102, 103, 109
Korea, 15, 83
Korean, 82
Korean War, 82

L

labor, 83, 85, 95
language, 37, 40, 88, 89, 107
law, viii, xi, 2, 3, 6, 8, 17, 35, 45, 58, 59, 61, 62, 65, 66, 67, 77, 78, 79, 80, 81, 83, 87, 94, 98, 100, 101, 105, 109, 111, 112, 115, 116, 117, 118, 121, 123, 125, 126
law enforcement, 3, 62, 78, 80, 87, 98
laws, vii, xi, 2, 5, 7, 62, 72, 89, 94, 100
lead, 2, 16, 19, 61, 75, 86
leadership, 37, 55, 88
learning, 21, 23, 28, 63
legal issues, xi, 72, 113
library services, 25, 28, 29
licenses, 118, 120
licensing, 101
light, 73, 85, 95, 116
limitations, 16
links, 104
litigation, 113, 115, 127
loan guarantees, 23, 27, 37, 46
loans, ix, 19, 21, 22, 23, 27, 28, 37, 43, 46, 53, 58, 59
local government, 42
location, 31, 119
logging, 102
long distance, 33
losses, 120
Louisiana, 105
lower-income, 55
low-income, viii, 11, 12, 14, 16, 20, 30, 33

M

Madison, 6, 9
magazines, 105
majority, 74, 83, 85
management, 4, 7, 32, 62, 78, 80, 94
management practices, 4, 7
mandates, 6, 42
manufacturer, 102
manufacturing, 114
mapping, 17, 60, 64, 65
market, 5, 6, 10, 15, 18, 19, 61, 63, 116, 121
market share, 5, 10
marketing, 102
marketplace, vii, ix, 1, 2, 5, 7, 9, 12, 17, 18, 46, 54, 67
markets, 16, 83, 120
Maryland, 28, 93
mass, 15, 75
Massachusetts, 15, 48
Massachusetts Institute of Technology, 15, 48
materials, 118, 119, 121, 124, 126
measures, ix, xii, 7, 12, 20, 46, 80, 81, 83, 86, 112, 116, 118, 124
media, 24, 47, 73, 77, 102
median, 14
medical, 24, 29
merger agreement, 4
mergers, 3
messages, 104
meta-tags, xii, 112, 116
Microsoft, 108, 109
military, x, 71, 72, 73, 74, 76, 81, 83, 86, 87, 88, 92, 94, 95, 97
Millennium, 115, 123
mirror, 50
mission, 21, 88
Mississippi, 26, 28
mobility, 51
modernization, 22
modifications, xii
mold, 84
money, 22, 23, 28, 39, 66
monopoly, 124
multimedia, 4
music, 119

N

nation, 16, 19, 33, 34, 37, 51, 54, 55, 67, 74, 75, 77, 78, 79, 82, 83, 84, 88, 89, 96
National Defense Authorization Act, 93
National Institutes of Health, 24, 29
National Intelligence Estimate, 89
national policy, 7, 9
National Science Foundation, 35
national security, x, 8, 71, 72, 73, 75, 76, 81, 82, 86, 89, 97
National Telecommunications and Information Administration, viii, ix, 2, 8, 22, 28, 37, 38, 48, 50, 53, 55, 68
National Telecommunications and Information Administration (NTIA), viii, ix, 2, 8, 53, 55
Native American, 25
Native Hawaiian, 25, 41, 42
negative consequences, 6
negative effects, viii, 2
negotiating, 121
net neutrality, vii, viii, 1, 2, 3, 4, 6, 7, 8, 42, 62
network, vii, viii, x, 1, 2, 3, 4, 5, 6, 7, 8, 10, 27, 33, 40, 42, 44, 54, 57, 62, 73, 74, 77, 78, 91, 92, 97
network interconnection obligations, viii, x, 2, 8, 40, 42, 44, 54, 57, 62
neutral, 4, 63, 64
New York, 28, 114, 122, 127
New York Times, 127
next generation, 13, 41, 63
nongovernmental, 41, 42
non-profit, 32
nonprofit organizations, 64
NTIA, viii, ix, x, 2, 8, 12, 17, 19, 38, 39, 40, 41, 42, 43, 44, 45, 50, 53, 54, 55, 56, 57, 58, 59, 60, 61, 62, 63, 64, 65, 66, 67

O

Obama, viii, ix, x, 12, 19, 20, 37, 43, 47, 50, 55, 72, 73, 75, 79, 90, 93
Obama Administration, ix, 12, 19, 73, 79
obligate, 86
obligation, 82
obligations, viii, x, 2, 8, 22, 23, 24, 25, 26, 40, 42, 44, 54, 57, 62, 66
OECD, 15, 48, 49
officials, 59, 77, 101
Ohio, 28
oil, 75
oil refineries, 75
Omnibus Appropriations Act,, 45
online, xii, 4, 66, 91, 103, 106, 107, 108, 109, 110, 112, 117, 120, 125
openness, 7, 44, 57, 62
operating system, 104
operations, x, 71, 77, 83, 84, 86, 97, 121
Organization for Economic Cooperation and Development, 15
originality, 125
oversight, ix, xi, 40, 43, 54, 55, 56, 66, 72, 76, 81, 86, 87, 88, 89, 97

P

parallel, 84
Paris, 127
participants, viii, 2, 8
partnership, 21, 26, 41, 42, 78, 109, 113, 125
partnerships, 17, 20, 45, 64, 80, 113
passive, 116
password, 102, 109, 110
personally identifiable information, xi, 99, 102, 105
personally identifiable information (PII), 102
Philadelphia, 125
photographs, 117, 124
pipelines, 75
planning, 22, 28, 38, 65, 81
play, viii, 12, 13, 19, 59, 63, 73
policy, ix, xi, 3, 4, 5, 6, 7, 8, 9, 12, 18, 19, 20, 28, 33, 46, 61, 62, 67, 72, 73, 79, 89
policy instruments, 46
policy makers, 33
policy responses, 79

policymakers, vii, viii, 1, 2, 5, 11, 12, 15, 19, 31, 46, 62, 65
political subdivision, 41, 42, 64
poor, 18
population, 14, 31, 38, 44, 57, 61
population density, 14
pop-up ads, 102, 103, 104
potential benefits, 5
poverty, 31
power, x, 5, 6, 16, 19, 72, 74, 75, 81, 82, 83, 85, 86, 87, 91, 93, 94, 95, 97
power lines, 6, 19
powers, x, 72, 80, 81, 82, 83, 85, 86, 89, 93, 94, 95, 96, 97
precedent, 115, 117, 118
President, viii, x, xi, 9, 10, 12, 18, 35, 36, 43, 45, 49, 50, 55, 71, 72, 73, 75, 76, 78, 80, 81, 82, 83, 84, 85, 86, 87, 88, 89, 90, 93, 94, 95, 96, 98, 101, 109
President Bush, 18, 78, 95
President Obama, viii, x, 12, 43, 55, 72, 73, 75, 90, 93
presidential authority, 86
presidential campaign, 73
Presidential Directives, 93
press, 48, 50, 73, 77, 90, 93, 122
prevention, 77, 78, 88
prima facie, xii, 112, 113, 115, 118
principles, vii, 1, 2, 3, 5, 6, 8, 9, 31, 33, 42, 44, 57, 62, 68, 85, 118, 126
privacy, 89, 101, 103, 108, 109, 110
private property, 83, 95
private sector, viii, ix, x, 11, 12, 16, 19, 28, 46, 54, 64, 67, 72, 73, 77, 78, 79, 80, 81
private sector investment, ix, 12, 19, 46, 54, 67
production, 83, 95
program, ix, x, 16, 21, 22, 23, 24, 25, 26, 27, 30, 31, 32, 34, 35, 36, 39, 40, 43, 44, 45, 50, 51, 54, 55, 56, 59, 60, 64, 67, 69, 77, 87, 90, 104, 106, 109
programming, 119
project, xii, 25, 27, 37, 40, 43, 44, 45, 57, 58, 112, 120
proliferation, 27
proposition, 97, 114, 118
prosperity, 15
Protect America Act, 94
protection, 8, 103, 106, 113, 119, 122, 126
protocol, 10
public, 3, 4, 8, 9, 10, 17, 18, 20, 21, 22, 28, 32, 33, 37, 39, 40, 41, 42, 43, 45, 46, 47, 49, 50, 51, 54, 55, 56, 57, 59, 60, 62, 64, 65, 66, 67, 68, 72, 73, 74, 78, 89, 91, 109, 117
public health, 74
public interest, 45, 57, 64
public notice, 4
public radio, 28
public safety, 28, 37, 39, 41, 55, 56
public television, 28
public welfare, 117
public-private partnerships, 17, 45, 64
publishers, xi, xii, 111, 112, 113, 114, 115, 125
publishing, xii, 112, 114

Q

quality of life, 25
quality of service, 4

R

radio, 28, 30, 124
random, 104
range, x, 4, 71, 72, 88, 98
rate of broadband deployment, viii, 11, 12, 14
REA, 21
recognition, 76
recommendations, iv, 64, 76, 90, 126
recovery, 68, 78, 88, 97
reform, vii, 1, 2, 6, 7, 19, 34, 46, 67
Reform, 50, 92
Reform Act, 92
reforms, 73, 80, 83
regional, 25, 27, 29
regulation, 2, 3, 6, 9, 54, 81, 85, 100, 101
regulations, viii, 2, 6, 8, 59, 60, 62
regulatory framework, 3, 7

regulatory guidelines, vii, 1, 5
relationship, 85
reproduction, xi, xii, 111, 112, 115, 117
requirements, 4, 6, 8, 42, 44, 56, 62, 66, 67, 87
researchers, 120, 124
reservation, 27
residential, 7, 13, 16
resolution, 36, 84, 87, 118, 124
resources, 20, 63, 74, 78, 98, 102
response, x, 34, 39, 71, 72, 73, 74, 76, 78, 84, 85, 86, 88, 89
responsibilities, 8, 88, 121
restrictions, vii, 1, 2
revenue, 126
Rhode Island, 27
rights, iv, xi, xii, 4, 9, 19, 72, 94, 96, 111, 112, 113, 114, 115, 116, 117, 118, 120, 121, 122, 124, 126
rights-of-way, 19
risk, 15, 63, 75, 76, 77, 78, 91
risk management, 78
risks, 79, 89, 92
routes, 83
routing, 4
rules, 32, 60, 62, 68, 81, 85
rural, viii, 11, 12, 14, 16, 20, 21, 22, 23, 24, 27, 28, 30, 31, 32, 33, 34, 35, 37, 40, 41, 43, 45, 47, 51, 55, 58, 61
rural areas, 14, 16, 21, 22, 23, 27, 28, 30, 41, 45, 47, 51, 61
Rural Broadband Access Loan and Loan Guarantee Program, 20, 23, 27, 30, 45
rural communities, 14, 20, 23, 27, 40
rural counties, 27
Rural Electrification Act, 21
Rural Utilities Service, ix, 20, 21, 22, 23, 27, 28, 30, 34, 35, 37, 40, 47, 51, 53, 55, 68
Rural Utilities Service (RUS), ix, 21, 34, 37, 53, 55
RUS, viii, ix, x, 12, 21, 27, 28, 30, 34, 36, 37, 38, 40, 41, 43, 45, 51, 53, 54, 55, 56, 58, 59, 60, 65, 66, 67, 68

S

safety, 28, 37, 39, 41, 55, 56, 74, 78, 109
satellite, viii, 6, 11, 13, 37, 39, 45, 54, 55, 64
scholarship, 116
school, 20, 23, 28, 31, 54
science, 24, 29
scope, ix, x, 34, 53, 55, 67, 72, 80, 81, 82, 84, 85, 86, 94, 117
search, xii, 22, 23, 24, 25, 26, 103, 109, 112, 113, 114, 116, 117, 121, 122, 125
search engine, xii, 112, 113, 114, 116, 117, 125
Secretary of Agriculture, 37, 40, 43, 58, 66
Secretary of Commerce, 40, 43, 56
security, x, 8, 71, 72, 73, 74, 75, 76, 77, 78, 79, 80, 81, 82, 84, 86, 88, 89, 92, 93, 97, 101, 104, 106
seed, 23, 28, 32
seizure, 83
selecting, 44, 59, 60
self-regulation, 100, 101
senate, 9, 49
Senate, xi, 7, 9, 10, 35, 36, 37, 39, 41, 42, 43, 45, 49, 55, 66, 75, 76, 87, 88, 91, 92, 100, 101, 105, 106, 107, 108, 109
separation, 81, 85, 86, 89
separation of powers, 89
series, viii, 11, 60
servers, 31
service provider, 3, 37, 55, 58, 62, 75, 78
services, iv, vii, viii, 1, 3, 4, 5, 6, 7, 8, 11, 19, 21, 24, 25, 28, 29, 30, 31, 32, 33, 34, 37, 38, 40, 43, 45, 46, 49, 51, 54, 62, 64, 67, 72, 75, 78, 91, 102
settlement class, xii, 112, 121, 127
shape, ix, 12, 20
sharing, 77
short-term, 67
signaling, 33
signs, xi, 100, 104
sites, 104
skills, 75
Small Business Act, 57

snippets, xi, xii, 111, 112, 113, 114, 115, 117, 120, 121
social consequences, viii, 11, 12
Social Security, 110
society, viii, 11, 12, 15, 118
software, xi, 99, 100, 101, 102, 103, 104, 105, 107, 108, 109
spam, 101, 104
spectrum, 19, 20, 46, 67
speech, 9, 31, 73, 90
speed, viii, 2, 8, 10, 11, 13, 14, 15, 16, 17, 18, 21, 28, 37, 40, 41, 42, 43, 44, 46, 49, 54, 57, 58, 63
spending, 37, 66, 87
spyware, vii, xi, 99, 100, 101, 102, 103, 104, 105, 106, 108, 109
stakeholders, ix, 54, 64
standards, 17, 19, 34, 36, 62, 65, 76, 78
state, 6, 14, 15, 16, 17, 25, 26, 28, 31, 33, 35, 36, 38, 40, 41, 42, 44, 54, 57, 59, 61, 63, 64, 65, 68, 72, 74, 75, 78, 79, 105
State Grants, 23, 28
state planning, 38
states, ix, x, xi, 17, 26, 28, 32, 34, 42, 50, 54, 58, 59, 60, 61, 63, 64, 68, 81, 100, 105
statistics, 48
statutes, 79, 80, 83, 84, 89
statutory, x, 72, 76, 77, 79, 80, 81, 82, 83, 86, 87, 89, 97, 117
statutory authority, 77
statutory provisions, x, 72, 82
steel, 82, 83
steel industry, 83
steel mill, 82
stimulus, viii, 2, 7, 12, 37, 46, 55, 67, 68
strategic planning, 28
strategies, 64, 77, 81
subscribers, viii, 8, 11, 12, 14, 17, 30, 121
subsidies, 19, 21, 27
suburban, 14, 15
supplemental, 81
supply, 78
supply chain, 78
Supreme Court, xi, 3, 8, 72, 81, 83, 84, 85, 86, 94, 97, 118
surfing, 4
surveillance, 82, 84
Sweden, 15
symptoms, 103

T

Taliban, 84
tanks, 37, 55
target, 30, 46, 97
tax credit, 19, 41, 42, 45
tax credits, 19, 42
tax incentive, 42, 46, 67
tax incentives, 42, 46, 67
taxes, 19
teaching, 23, 116
technical assistance, 33, 64
technological advancement, 101
technologies, viii, 6, 11, 12, 13, 19, 21, 34, 47, 54, 63, 77, 100, 101, 103
technology, 14, 19, 20, 23, 24, 28, 29, 41, 46, 47, 50, 54, 63, 75, 77, 78, 82, 103, 106
technology gap, 28
telecommunication, 23, 74, 79, 86
telecommunications, vii, viii, x, 1, 2, 3, 6, 7, 11, 12, 14, 15, 16, 18, 20, 21, 22, 23, 28, 30, 31, 32, 33, 34, 37, 54, 55, 72, 74, 75, 77, 78, 84, 89, 91, 97, 98
Telecommunications Act, 18, 31, 33, 49, 50
telecommunications reform, vii, 1, 2, 6, 7
telecommunications services, viii, 3, 11, 21, 30, 31, 32, 33, 34, 54
telehealth, 24, 28
telemedicine, 21, 63
telephone, viii, 3, 6, 7, 11, 13, 21, 22, 27, 30, 38, 43, 54, 58, 124
telephone companies, 3, 13
telephones, viii, 11, 12
television, 3, 7, 13, 28, 118, 119
Tennessee, 105
territory, 44, 57, 64
terrorism, 73, 84, 96
terrorist, x, 71, 76, 80, 84

terrorist attack, 76, 80
terrorist groups, x, 71, 76
terrorist organization, 84
testimony, 9, 10, 73, 101, 105, 110
textbooks, 121
theft, 102, 109
threat, x, 71, 74, 75, 78, 79, 86, 101
threats, x, 71, 72, 73, 74, 76, 77, 78, 79, 80, 82, 83, 86, 87, 88, 89, 97, 103
threshold, 47, 63
thresholds, 44, 57, 63
thumbnails, 118, 124
time frame, ix, 53, 55
Title I, 3, 5, 8, 37, 39, 50, 92
Title II, 3, 8, 37, 92
Title III, 37, 92
Title V, 36, 37, 44, 56
tracking, 59, 102
training, 12, 20, 33, 39, 56
transcript, 110
transfer, 41, 114
transition, 19, 37, 55, 74
transparency, x, 54, 66, 67, 78, 88, 101
transparent, 79

U

U.S. Department of Agriculture, ix, 20, 27, 28, 35, 53, 55
U.S. Department of Agriculture (USDA), ix, 53, 55
U.S. Department of Commerce, 48, 50
U.S. military, 73, 92
U.S. policy, 31
unauthorized digitization, xii, 112, 115, 120
uncertainty, 16, 33
unclassified, 77, 88, 89
upload, 13, 17, 41, 42, 61, 63
urban, viii, 11, 12, 14, 15, 32, 33, 48
urban areas, 14, 15, 32, 33
USA, 90
US-CERT, 91
USDA, viii, ix, 12, 30, 35, 40, 43, 45, 47, 51, 53, 55, 56, 58, 68
Utah, 105, 109

V

values, 120
vehicles, 88
venture capital, 32
voice, 4, 10, 13, 17, 27, 33, 38, 51, 76
voice over Internet Protocol, 17
VoIP, 10, 17, 49
vulnerability, 73, 74, 75

W

wants and needs, 16
war, x, 72, 81, 82, 83, 84, 86, 94, 95, 96, 97
war on terror, 84, 95, 96
War on Terror, 84
War Powers Resolution, 86, 96, 97
warfare, 81
Washington, 68
waste, 66
weapons, 75, 85
weapons of mass destruction, 75
web, 101
websites, 68, 97, 100
White House, 50, 76, 90, 91, 93
windows, 104, 107
wireless, viii, 6, 11, 13, 17, 19, 27, 35, 37, 38, 39, 41, 42, 44, 45, 46, 51, 54, 55, 57, 60, 61, 64, 67
wireless systems, 13, 54
wiretapping, 109